MATHEMATICS
FOR DATA
PROCESSING

MATHEMATICS FOR DATA PROCESSING

Second Edition

Frank J. Clark

Reston Publishing Company, Inc.
A Prentice-Hall Company
Reston, Virginia 22090

Library of Congress Cataloging in Publication Data

Clark, Frank James
  Mathematics for data processing.

  Includes index.
  1. Mathematics—1961–   . I. Title
QA39.2.C57 1982   510'.2409   82-15094
ISBN 0-8359-4263-5

Interior design/production: Jeanne-Marie Peterson

© 1983 by Reston Publishing Company, Inc.
  A Prentice-Hall Company
  Reston, Virginia 22090

10   9   8   7   6   5   4   3   2   1

Printed in the United States of America

CONTENTS

PREFACE

With computers now several "generations" old and in use in widely diversified business and commercial applications, it is becoming easier to describe the mathematical background for the student seeking an entry occupation in data processing.

This mathematical background can probably be best described in terms of the expectations of future employers of these students. Students for entry occupations in data processing should:

Have a thorough foundation in the characteristics of the decimal system and the properties of a "field."

Understand the representation of data internally in computers.

Know how the computers perform arithmetic.

Be skilled in arithmetic operations in the binary, octal, and hexadecimal numeration systems.

Understand the logic of networks, Boolean algebra, and some Set Theory.

As Algebraic methods are being used more frequently in the solution of business problems, it follows that a data processing mathematics course should provide students with a strong foundation in the following areas:

Linear Functions
Nonlinear Functions
Logarithmic Functions

Simultaneous Systems of Equations
Matrices
Linear Programming

Statistical measurements also are commonly used as a method of summarizing data, predicting future trends, or providing management with a tool for decision making. Students of data processing should not only possess a vocabulary in this area, but also know how to perform and write simple programs for:

Measures of Central Tendency
Measures of Dispersion

Some high school mathematics topics are reviewed in Appendix B. The instructor may wish to review them before proceeding to Chapter 7.

In some sections of the text, emphasis is placed on the mathematics used in solving problems for computer applications and not on rigorous proofs. The purpose and scope of *Mathematics for Data Processing* is to provide students with necessary and sufficient skills so that they will be proficient in their area when they enter the data processing field.

Finally, I'd like to express my thanks and appreciation for the helpful suggestions of my colleagues, Dr. Peter Lindstrom, Dr. Bernard Hoerbelt, and Jim Hofmann.

0

INTRODUCTION

0-1 SETS AND NUMBER SYSTEMS

In a discussion of computer mathematics an understanding of the various kinds of number systems is essential. We will use sets to aid us in our study of number systems. A set is a collection of objects and is usually denoted by a capital letter. A collection may include such things as people, inanimate objects, numbers, or anything else. Any one item in a collection or set is called a *member* or an *element* of that set. Elements in a set are usually enclosed in brackets, { }.

0-1.1 Set Notation

In discussing numbers and number systems, we will use set notation. The elements in a set may be represented in the *listing format* as follows:

$$\text{set } A = \{\triangle, \bigcirc, \square\}$$

In this case set A is a collection containing a triangle, a circle, and a square. Elements, or members, are listed within brackets and separated by commas. To denote $\triangle$ as an element of A, we write $\triangle \in A$. To indicate that an element is not a member of a set, we write $\star \notin A$ where $\star$, a star, is not a member of the set A.

Lower-case alphabetic characters (e.g., a, b, c, x, y, etc.) are frequently used to indicate the elements in a set.

Another way of representing the elements in a set is called *set builder* notation:

$$C = \{x \mid x \text{ is any natural number and } x > 10\}$$

This means that set C contains every x such that x is any natural number greater than 10. The vertical line is read "such that." If we had wanted to write set C in the listing format mentioned previously, we could have written: $C = \{11, 12, 13, \ldots\}$. The three trailing dots denote that the sequence continues indefinitely.

0-1.2 Number Classification

Natural numbers consist of counting numbers. The set of natural numbers can be written as $N = \{1, 2, 3, \ldots\}$.

Integers are a set of numbers that includes the natural numbers, their negatives, and zero. $J = \{\ldots -3, -2, -1, 0, 1, 2, 3, \ldots\}$. Note that N is included in J; i.e., each element of N is also an element of J. We can express the integers in set builder notation as: $J = \{a \mid a = -b, 0, \text{ or } b, \text{ where } b \in N\}$.

Rational numbers are those numbers that can be represented as a fraction or a quotient of two integers x/y, where $y \neq 0$.

An example of a rational number is $\frac{1}{2} = 0.5000$, where any expansion after 0.5 yields zero. Similarly, $\frac{1}{4} = 0.2500$. Rational numbers that terminate, or yield zero after a finite number of steps, are called *terminating decimals*. Consider other examples where expansions never terminate but repeat themselves in groups as follows:

$$\frac{2}{3} = 0.666 \ldots$$
$$\frac{1}{7} = 0.142857142857 \ldots$$

When the digit or group of digits is repeated, the decimal is referred to as a *repeating decimal expansion*.

Terminating decimals and repeating decimals belong to the set of rational numbers that can be represented as x/y where $y \neq 0$. The symbols x and y are used to represent any value and are called *variables*. In this case x and y are integers and y is not equal to zero. We can express rational numbers in set builder notation as $Q = \{x/y \mid x, y \in J \text{ and } y \neq 0\}$.

Irrational numbers are those numbers that are not rational. Such numbers would be decimal numbers that do not terminate and do not repeat. Some examples of irrational numbers are $\sqrt{2}, \pi, \sqrt{7}$. Note that the numbers that are rational and the numbers that are irrational form two sets with no elements in common. Such sets which have no elements in common are said to be *disjoint*. We can express irrational numbers in set builder notation as

$$H = \{h \,|\, h \text{ represents a decimal expansion that}$$
$$\text{does not repeat and does not terminate}\}$$

Examples are $\sqrt{2} = 1.4142 \ldots$, and $\pi = 3.1415. \ldots$

Real numbers include all of the rational and irrational numbers and can be expressed as

$$R = \{x \,|\, x \in Q \text{ or } x \in H\}$$

The sets of numbers are related as shown in the diagram below.

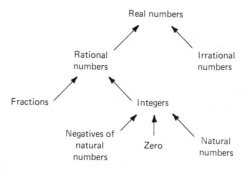

FIGURE 0-1

There are other sets of numbers that are not usually found in business data processing problems. They are the set of complex numbers expressed in the form $x + yi$ where $x \in R, y \in R$ and $i = \sqrt{-1}$. If $x = 0$, the complex number $x + yi$ can be written as yi and denotes a pure imaginary number. If $y = 0$, the imaginary term drops out, and in this special case the number $x + yi$ is a real number.

We will not discuss imaginary and complex numbers any further because they are not part of typical business data processing problems.

Exercises 0-1

1. Write the numbers of each set, using the listing format. Use three trailing dots (. . .) where appropriate.

   a. Natural counting numbers from 3 to 8
   b. Integers less than 4
   c. Integers between -20 and -15
   d. Natural numbers less than 4
   e. Nonnegative integers
   f. Natural numbers greater than 7

2. Let $X = \{3, \sqrt{2}, -4, 6, \frac{1}{2}, -8, \pi\}$. Denote with braces the following sets, using the listing method.

   a. Natural numbers in X
   b. Real numbers in X

    c. Integers in X

    d. Rational numbers in X

3. Use the set-builder format, $\{n \mid \text{the conditions of } n\}$, to represent each of the following sets.

    a. Real numbers

    b. Natural numbers greater than 6

    c. Rational numbers between $\frac{1}{4}$ and $\frac{5}{2}$

    d. Nonnegative rational numbers

    e. Integers less than zero

0-2 A FIELD

A field is an abstract system in mathematics containing elements, operations, relations, and axioms.

At least two elements are assumed to exist, and operations are described as binary operations. Operations may be addition, denoted by $+$, and multiplication denoted by $\times$. However, in some systems, operations need not be addition or multiplication. A binary operation manipulates and assigns to a pair of elements the element which is formed as a result of the binary operation. Binary operations are used in simple arithmetic, although the term "binary" is seldom used. A simple example is $1 + 2 = 3$, where the pair of elements is (1, 2) and the resulting element is 3. The result of a binary operation may not always be a *new* element, as for example, in $0 + 0 = 0$.

The equals, or equality, relation is described by the following laws:

F-1	$a = a$	(Reflexive Law)
F-2	If $a = b$, then $b = a$	(Symmetric Law)
F-3	If $a = b$ and $b = c$, then $a = c$	(Transitive Law)
F-4	If $a = b$, then a may be replaced by b and b may be replaced by a	(Substitution Law)

Axioms describe the behavior of the elements under binary operations. Throughout this book, unless otherwise stated, the elements will be real numbers.

0-2.1 Axioms of Operations

The following hold true for all real numbers a, b and c:

1 CLOSURE AXIOM FOR ADDITION

$$a + b \text{ yields a unique real number}$$

(In the language of sets, if $a \in R$ and $b \in R$, then $a + b \in R$)

0-2.6   Division

The operation of finding a quotient of two numbers is called division. In Axiom 10 we expressed the operation of division in the term $1/a$. If we denote the multiplicative inverse of a as a^{-1}, then $a \times 1/a = a \times a^{-1} = 1$.

This nomenclature for the multiplicative inverse makes it possible for us to express the quotient of two numbers a/b as $a \times b^{-1}$ and is consistent with the axioms given for a field. We can define division by stating that a number x is said to be divided by a number d if there exists a number q such that $q \cdot d = x$. The number d is called the divisor, and the number q is called the quotient. The number x is called the dividend.

Exercises 0-2

1. State the law (F-1, F-2, F-3, F-4, or F-5) that applies to each of the following.

   a. $x = 4, 4 = x$
   b. If $a = b + 3$ and $b = 2$, then $a = 2 + 3$
   c. If $a - 2 = b + 3$ and $b + 3 = x^2$, then $a - 2 = x^2$
   d. If $a \in R$, then $a = 0$, or $a > 0$, or $a < 0$

2. State the axioms appropriate to each of the following.

   a. $(x + 3) + (-b) = x + [3 + (-b)]$
   b. $(x + y) \in R$
   c. $x + ya + yb = x + y(a + b)$
   d. $1 \cdot a = a$
   e. $4(3 + 2) = (3 + 2)4$
   f. $xy = yx$
   g. $2a + 3a = a2 + a3$
   h. $(1 + m)\dfrac{x}{y} = \dfrac{x1}{y} + \dfrac{xm}{y}$
   i. $-b + 0 = -b$
   j. $cy + ab = ab + cy$
   k. $4[(5 + 3) + 2] = 4(5 + 3) + 4 \cdot 2$
   l. $\dfrac{a}{x + 2} \cdot \dfrac{x + 2}{a} = 1,\ (x + 2 \neq 0)$
   m. $\dfrac{3}{4} + \dfrac{2}{4} = \dfrac{2}{4} + \dfrac{3}{4}$
   n. $(ab)c = a(bc)$
   o. $a + -a = 0 = -a + a$
   p. $a^{-1} \cdot a = 1$

3. Is $\{(x + 2)|x \in N\}$ closed* under addition? (N is the set of natural numbers.)
4. Is $\{0, 1, -1\}$ closed under multiplication?
5. Is R (the set of real numbers) closed under division?
6. Is $\{2n|n \in Q\}$ closed under addition? (Q is the set of rational numbers.)
7. Are natural numbers closed under subtraction?
8. Are natural odd numbers closed under division?
9. Are nonnegative integers closed under division?
10. Is subtraction commutative for real numbers?
11. Is division commutative for irrational numbers?
12. Given the undefined operation $\ne$, the set of elements $\{*, !, \triangle, /\!/\}$, and the following:

$\ne$	$*$	$!$	$\triangle$	$/\!/$
$*$	$!$	$\triangle$	$/\!/$	$*$
$!$	$\triangle$	$/\!/$	$*$	$!$
$\triangle$	$/\!/$	$*$	$!$	$\triangle$
$/\!/$	$*$	$!$	$\triangle$	$/\!/$

FIGURE 0-7

a. Is the set of symbols closed with respect to $\ne$?
b. Is the operation commutative?
c. Is the operation associative?
d. Is there an identity element?
e. Is there an inverse element for each of the given elements?

13. Given the number line

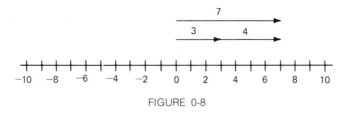

FIGURE 0-8

and the graphic representation of $3 + 4 = 7$ as a model, illustrate

---

* A set is *closed* under an operation if and only if, whenever the operation is applied to the appropriate element(s) of the set, the result is an element of that set.

  a. $6 + 2 = 8$
  b. $10 - 3 = 7$
  c. $4 + 2 = 2 + 4$
  d. $5 - 3 = 2$
  e. $(1 + 2) + 3 = 3 + (2 + 1)$
  f. $2 \cdot 3 = 6$
  g. $2(3 + 1) = 2 \cdot 3 + 2 \cdot 1$
  h. $2 \cdot 3 = 3 \cdot 2$

14. Find the absolute value of

  a. 7
  b. 3/7
  c. $-.73$
  d. 0
  e. $x - 2$, where $x - 2$ is greater than or equal to zero

0-3 FIELD PROPERTIES

The laws and axioms imply properties of a field. These properties can be expressed as theorems. Theorems usually are expressed by an "if . . . then" statement. The *if* part of the theorem is called the hypothesis, and the *then* part is called the *conclusion*. Proving these theorems requires showing that a conclusion is a logical result of the laws of the system. Field properties describe the behavior of the elements with respect to the operations. One such property is the Addition Law for Equality, which states, in a very general way, that equals added to equals are equal. More specifically:

If $a \in R$, $b \in R$, $c \in R$, and $a = b$, then $a + c = b + c$

The Multiplication Law for Equality states that equals multiplied by equals are equal:

If $a \in R$, $b \in R$ and $a = b$ then $ac = bc$, when $c \in R$

Theorem 0-1

If $a = b$, then $a + c = b + c$

Proof:

By the Reflexive Law (*F*-1)

(A) $a + c = a + c$

but $a = b$, so that (A) becomes $a + c = b + c$ by the Substitution Law (*F*-4)

Theorem 0-2

Cancellation Law of Addition
If $a \in R$, $b \in R$, and $c \in R$, and $a + c = b + c$, then $a = b$
Proof:

Since $a + c = b + c$, by Theorem 0-1:
$$(a + c) + (-c) = (b + c) + (-c)$$
$$a + [(c + (-c)] = b + [(c + (-c)] \quad \text{Associative axiom for addition (3)}$$
$$a + 0 = b + 0 \quad \text{Inverse axiom for addition (5)}$$
$$a = b \quad \text{Identity axiom for addition (4)}$$

Theorem 0-3

Given $a \in R$, $b \in R$, $c \in R$.
If $a = b$, then $a \cdot c = b \cdot c$
Proof:

By the Reflexive Law (F-1)
(B) $a \cdot c = a \cdot c$
But $a = b$ so that (B) becomes $a \cdot c = b \cdot c$ by the Substitution Law
(F-4)

Theorem 0-4

Cancellation Law for Multiplication
If $a \in R$, $b \in R$, $c \in R$, and $ac = bc$, then $a = b$, when $c \neq 0$.
Proof:

Since $ac = bc$, by theorem 0-3:
$$(ac) \cdot 1/c = (bc) \cdot 1/c$$
$$a(c \cdot 1/c) = b(c \cdot 1/c) \quad \text{Associative axiom for multiplication (8)}$$
$$a(1) = b(1) \quad \text{Inverse axiom for multiplication (10)}$$
$$a = b \quad \text{Identity axiom for multiplication (9)}$$

All of the remaining important properties of a field will be noted in the exercises that follow. The student should be familiar with all of them and be able to describe them verbally as well as express them in algebraic form.

Exercises 0-3

1. Write a sentence describing each of the following.
   a. If $a \in R$, $b \in R$, and $a + b = 0$, then $a = -b$ and $b = -a$
   b. If $a \in R$, $b \in R$, and $a \cdot b = 0$, then $a = 0$ or $b = 0$, or both a and b = 0
   c. $-(-a) = a$ (where $a \in R$)

d. $-a + -b = -(a + b)$ (where $a, b \in R$)

e. $(-a)b = -(ab)$ (where $a, b \in R$)

f. $\dfrac{-x}{-y} = \dfrac{x}{y}$ ($y \neq 0$, $x, y \in R$)

g. $\dfrac{a}{b} = \dfrac{c}{d}$ if and only if $ad = bc$ (where $a, b, c, d \in R$)

2. Write the algebraic expression for each of the following.

   a. 1 divided by the multiplicative inverse of a real number yields the real number.

   b. A negative real number multiplied by another negative real number yields a positive real number.

   c. A positive real number divided by a negative real number yields a negative real number, and a negative real number divided by a positive real number yields a negative real number.

   d. Every real number multiplied by zero yields zero as a result.

3. The following properties should be remembered from introductory algebra. Write a sentence to describe each one. Each variable (a, b, x, y) represents a real number.

   a. $\dfrac{ax}{bx} = \dfrac{a}{b}$ ($b, x \neq 0$)

   b. $\dfrac{1}{x} \cdot \dfrac{1}{y} = \dfrac{1}{xy}$ ($x, y \neq 0$)

   c. $\dfrac{a}{b} \cdot \dfrac{x}{y} = \dfrac{ax}{by}$ ($b, y \neq 0$)

   d. $\dfrac{a}{x} + \dfrac{b}{x} = \dfrac{a + b}{x}$ ($x \neq 0$)

   e. $\dfrac{a}{x} - \dfrac{b}{x} = \dfrac{a - b}{x}$ ($x \neq 0$)

   f. $\dfrac{x}{a} + \dfrac{y}{b} = \dfrac{xb - ay}{ab}$ ($a, b \neq 0$)

   g. $\dfrac{x}{a} \cdot \dfrac{y}{b} = \dfrac{xb + ay}{ab}$ ($a, b \neq 0$)

   h. $\dfrac{1}{\frac{a}{b}} = \dfrac{b}{a}$ ($a, b \neq 0$)

   i. $\dfrac{\frac{a}{b}}{\frac{x}{y}} = \dfrac{ay}{bx}$ ($b, y, x \neq 0$)

0-4 COMPUTER DATA ITEMS

Computer programming involves computations using such items of data as signed numbers, constants, variables, terms, and expressions.

Definitions and Nomenclature

Definition 0-1

A *signed number* is a number that is preceded by a plus or minus sign.

EXAMPLE 1   A *positive number* is signed plus (e.g., $+2$, $+3$), or considered positive if no sign appears in front of it (e.g., 7, 4).

EXAMPLE 2   A *negative number* is a number that is preceded by a minus sign (e.g., -1, -5).

Definition 0-2

A *constant* is a value that remains unchanged throughout a given procedure (e.g., 4, -32, 0.071).

Variables are discussed on page 2 of this chapter. Typically an alphabetic character, a variable has different values under different conditions.

Definition 0-3

A *term* consists of constants, variables, or combinations of either or both. Constants or variables combined by multiplication are terms. (By the identity axiom for multiplication, x may also be considered as $1x$.)

EXAMPLE 3 $3s$, $4y$, $2xabc$, $-8x^2y^3$, y are terms. ($1 \cdot y$ is implied in y.)

Definition 0-4

An *expression* consists of constants or variables combined by any of the arithmetic operators. An expression must contain at least one constant.

EXAMPLE 4 $2x - 1$, $x^2 + 3x + C$, $(x + 1)$, $2/3x^3$, x/y are expressions.

In such programming languages as Cobol, Basic, Fortran and PL/1, variables may contain more than one character. This is in contrast to algebra, where each alphabetic character represents a different variable. Variables may also include numeric characters.

Typical variable names are: X, Y, X1, Y346, MASS, ANSWER.

Words or letters are always written in upper case (capitals).

Being able to use more than one character in a variable name in computer programming provides for meaningful flexibility when writing expressions.

In Fortran, two types of arithmetic modes are possible:

1. Fixed-point arithmetic
2. Floating-point arithmetic

Fixed-point arithmetic is concerned with operations using integers such as 1, − 65340, etc.

Floating-point arithmetic is concerned with operations using real numbers such as 7.30, − 0.002, 3.333, etc.

All variable names must begin with an alphabetic character. Variables in fixed-point arithmetic must begin with one of the letters I, J, K, L, M, or N. Variables names in floating-point arithmetic must begin with any letter *except* I, J, K, L, M, or N.

0-4.1 Flowcharts

To use a computer to solve a problem, the information must be presented in an orderly way. The problem must be reasoned out from input through processing to final output in a series of sequential and documented steps. Any exceptions in these sequential steps must be identified. A flowchart is the most common method of reasoning out and documenting these sequential steps. Flowcharts consist of symbols, annotations, and flowlines. (See Fig. 0-9.)

A *flowchart* is a graphic description of the logic used to solve a problem. Flowcharts present computer operations, including decision making, in a visual manner that is easy to follow.

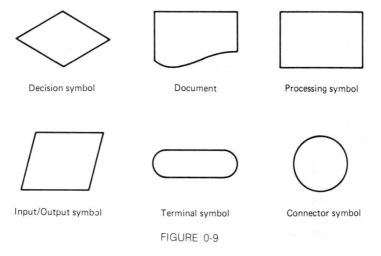

Decision symbol   Document   Processing symbol

Input/Output symbol   Terminal symbol   Connector symbol

FIGURE 0-9

Symbols represent a function, i.e., an arithmetic operation, a decision, a modification, and so on.

Annotations within symbols describe the function, i.e., addition, test for a negative number, and so on.

Flowlines connect the symbols and indicate the direction of sequential steps.

The flowchart symbols used in this text conform to revised flowchart symbols found in the International Organization for Standardization (ISO) Draft Recommendation on Flowchart Symbols for Information Processing. They are consistent with the set of fewer symbols adopted by the U.S.A. Standards Institute (USASI).

Consider a situation where real numbers on punched cards are being read and only the positive and nonzero values are added to an accumulator or register and printed.

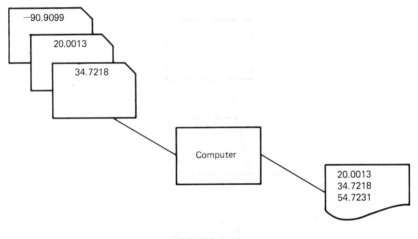

FIGURE 0-10

Figure 0-10 shows a pictorial representation of the activities mentioned above. Figure 0-11 is a flowchart detailing the required sequence of activities. Note that the flowchart shows us clearly all the steps involved. These steps include: reading the cards, adding only the positive numbers, printing the positive numbers, and printing the total of the positive numbers.

The shape of the decision symbol makes possible a choice among three alternatives. In Fig. 0-11 we only tested for two alternatives, but could have tested the number for positive, negative, or zero values and then proceeded to one of three steps depending upon the results of the test. This type of decision operation in shown in Fig. 0-12.

The normal flow of information in a flowchart is from top to bottom and from left to right. When this flow changes from bottom to top or from

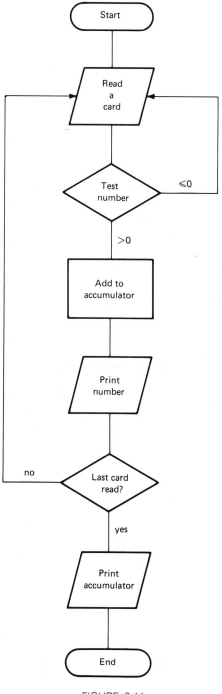

FIGURE 0-11

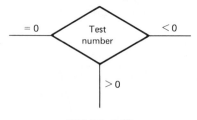

FIGURE 0-12

right to left, the flowlines contain arrowheads. Note the arrowheads in Fig. 0-11 when the flow changes from bottom to top.

Exercises 0-4

1. A and B are real numbers found on punched cards. Use flowlines and place the following symbols in the appropriate places to arrange the values A and B in ascending numerical order and print out the result.

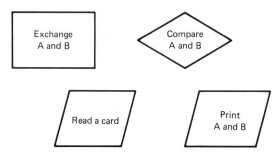

FIGURE 0-13

2. Draw a flowchart to read and print negative numbers only.
3. Identify the following constants and variables as fixed point or floating point (in Fortran).
   a. 0.666
   b. 666.0
   c. −123
   d. GROSS
   e. NET
   f. VOLTS
   g. K80
   h. ANSWER
4. The Dillon retail store has the following charge account plan:

   $11 is the minimum monthly payment. If a customer's balance is greater than $11, bill the customer for ⅓ of the current balance.

If a customer's balance is equal to or less than $11, the entire amount is due.

If a customer's balance is zero for any given month, the customer receives no bill for that month.

Flowchart the rules for payment in the Dillon charge account plan.

1

NUMERATION SYSTEMS

Communications between man and the computer are similar to communications between persons. For communication we need a set of symbols whose meanings we all agree on. These symbols are the medium through which we exchange information.

1-1 THE DECIMAL SYSTEM AND MECHANICAL COUNTING

When we wish to provide answers to such questions as "how much" or "how many," we usually use the set of symbols that are members of the decimal system. The most widely used number system is the decimal system with a base of ten. One explanation of why we use a system based on ten is that man made a one-to-one correspondence between the objects he wanted to count and his fingers. However, there are number systems with bases other than ten, and we shall discuss them later.

As commerce increased, crude computational devices were invented. A sand counter is shown in Fig. 1-1, and an abacus is shown in Fig. 1-2.

These primitive devices gave *quick and dependable* answers and were the beginning of man's search for mechanical aids in computation.

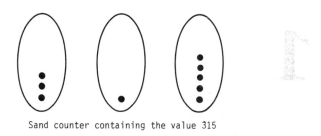

Sand counter containing the value 315

FIGURE 1-1

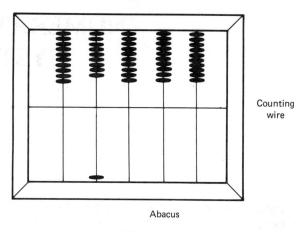

Counting wire

Abacus

FIGURE 1-2

Note that ten beads are found on each wire of the abacus. Nine of the beads can be denoted by the digits 1-9. As soon as the tenth bead is pulled below the counting wire, these ten beads must be pushed above that wire and exchanged for one bead in the next column to the left. Thus we get our symbol for 10. A wire containing no beads below the counting wire is denoted by the symbol 0.

In the thirteenth century, clocks using wheels and gears were able to measure time, but it took another 400 years before someone was able to see that wheels and gears could also count "how many."

In 1642 Blaise Pascal applied the concept of gears to an automatic adding device. Imagine the wire of the abacus now as forming a circle, and the ten beads on the abacus as ten teeth on a counting wheel. As soon as ten has been counted in any wheel, a carry is forced into the next wheel to the left. Desk calculators, adding machines, odometers, and cash registers are all examples of how the decimal system has been "fitted" onto man's mechanical devices.

Between these mechanical counters and today's computer are the Unit Record machines. (See Fig. 1-3, an IBM accounting machine.) Some of these Unit Record machines performed many of the same func-

FIGURE 1-3

tions as desk calculators. They did their computations in the decimal system by electrically driven counting wheels.

Data enters Unit Record equipment through the punched card, the way that it still enters many computers. The punched holes in the card behave like an electrical switch. This method is illustrated in Fig. 1-4.

A punched hole allows the brushes to complete a circuit. Where no hole is punched, no circuit is completed.

The code most frequently used with cards is the Hollerith Code. This

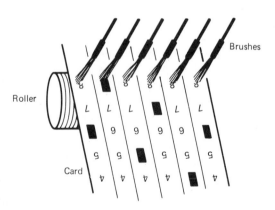

FIGURE 1-4 From *Computers* by Vorwald and Clark. Used with permission of McGraw-Hill Book Company.

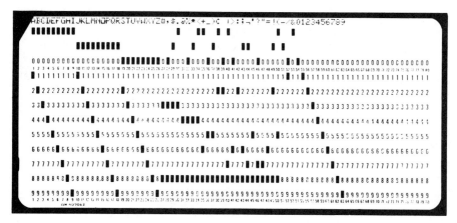

FIGURE 1-5

code was named for its creator, Dr. Herman Hollerith, a statistician for the Census Bureau in 1890.

Figure 1-5 shows a card containing various combinations of punches used in the Hollerith Code.

Number Symbols

To record their computations, early peoples invented number symbols. (See Fig. 1-6.)

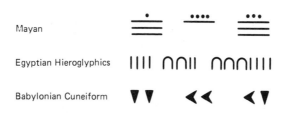

FIGURE 1-6

Early counting number symbols were cumbersome, and hindered mathematical thought for thousands of years. With the invention of zero, which may be thought of as the written placeholder for an empty wire on an abacus, mathematical thought was revolutionized and freed from the use of so many different symbols.

Using zero, man would write any number, large or small, with a combination of ten basic symbols 0, 1, 2, 3, 4, 5, 6, 7, 8, and 9. This greatly simplified the mathematical process.

Writing numbers in the decimal system of numeration means giving numbers a *face* value and a *place* value. Each of the ten symbols in the system has both a face and a place value.

The number 555 contains three symbols each with the same *face* value. Because each symbol 5 represents a different power of base ten, each of the three 5's has its own *place* value. It was the symbol zero that made this concept of place value possible.

The number 555 expressed in powers of base ten can be written as

$$5 \times 10^0 = 5$$
$$5 \times 10^1 = 50$$
$$5 \times 10^2 = 500$$

and $5 + 50 + 500 = 555$. Refer to the abacus in Fig. 1-2. It is clear that each successive place to the left contains a value that is ten times greater than the previous one. Expressing these values in powers of base ten, we find:

First Wire	$10^0 = 1$
Second Wire	$10^1 = 10$
Third Wire	$10^2 = (10)(10) = 100$
Fourth Wire	$10^3 = (10)(10)(10) = 1000$
Fifth Wire	$10^4 = (10)(10)(10)(10) = 10,000$

The number 9726 expressed in terms of the powers of base ten is

$$6 \times 10^0 = 6$$
$$2 \times 10^1 = 20$$
$$7 \times 10^2 = 700$$
$$9 \times 10^3 = 9000$$

where $6 + 20 + 700 + 9000 = 9726$

The 6 is called the units digit, 2 is the tens digit, 7 the hundreds digit, and 9 the thousands digit. Visualizing the number 9726 on the abacus, we think of the 9 as representing 9 groups of 1000, the 7 as seven groups of 100, the 2 as two groups of 10, and the 6 as six groups of 1.

1-1.1 Expanded Notation

To understand the operations of addition, subtraction, multiplication, and division of numbers in the decimal system, a clear understanding of place value is required. The following examples are written in *expanded notation* and illustrate this concept.

Consider the number

$$\ldots a_2\ a_1\ a_0$$

written in expanded notation as

$$\ldots + a_2(10^2) + a_1(10^1) + a_0(10^0)$$

Using this general format of expanded notation, we can now write the decimal number 493 as

$$493 = 4(10^2) + 9(10^1) + 3(10^0)$$
or
$$493 = 4(10)^2 + 9(10) + 3$$

We can perform all of the basic arithmetic operations in expanded notation, as shown in the following examples:

ADDITION

$$
\begin{array}{r}
25 \\
+\ \ 12 \\
\hline
37
\end{array}
\qquad \text{means} \qquad
\begin{array}{r}
2(10) + 5 \\
+\ 1(10) + 2 \\
\hline
3(10) + 7
\end{array}
$$

$$
\begin{array}{r}
365 \\
+\ 278 \\
\hline
643
\end{array}
\qquad \text{means} \qquad
\begin{array}{r}
3(10)^2 +\ \ 6(10) +\ \ 5 \\
+\ 2(10)^2 +\ \ 7(10) +\ \ 8 \\
\hline
5(10)^2 + 13(10) + 13 \\
=\ 6(10)^2 +\ \ 4(10) +\ \ 3
\end{array}
$$

(note the carries into the tens and hundreds columns)

SUBTRACTION

$$
\begin{array}{r}
63 \\
-\ \ 41 \\
\hline
22
\end{array}
\qquad \text{means} \qquad
\begin{array}{r}
6(10) + 3 \\
-\ 4(10) + 1 \\
\hline
2(10) + 2
\end{array}
$$

$$
\begin{array}{r}
463 \\
-\ 174 \\
\hline
289
\end{array}
\qquad \text{means} \qquad
\begin{array}{r}
4(10)^2 +\ \ 6(10) +\ \ 3 \\
-\ 1(10)^2 +\ \ 7(10) +\ \ 4
\end{array}
$$

we can rearrange or rename our minuend to indicate "borrowing" as follows:

$$
\begin{array}{r}
3(10)^2 + 15(10) + 13 \\
-\ 1(10)^2 +\ \ 7(10) +\ \ 4 \\
\hline
2(10)^2 +\ \ 8(10) +\ \ 9
\end{array}
$$

MULTIPLICATION

$$
\begin{array}{r}
22 \\
\times\ \ 11 \\
\hline
22 \\
22\ \ \\
\hline
242
\end{array}
\qquad \text{means} \qquad
\begin{array}{r}
2(10) +\ \ 2 \\
\times\ \quad 1(10) +\ \ 1 \\
\hline
2(10) +\ \ 2 \\
2(10)^2 +\ \ 2(10) \\
\hline
2(10)^2 +\ \ 4(10) +\ \ 2
\end{array}
$$

$$
\begin{array}{r}
32 \\
\times\ \ 18 \\
\hline
256 \\
32\ \ \\
\hline
576
\end{array}
\qquad \text{means} \qquad
\begin{array}{r}
3(10) +\ \ 2 \\
\times\ \quad 1(10) +\ \ 8 \\
\hline
24(10) + 16 \\
3(10)^2 +\ \ 2(10) \\
\hline
3(10)^2 + 26(10) + 16 \\
=\ 5(10)^2 +\ \ 7(10) +\ \ 6
\end{array}
$$

DIVISION

$$\begin{array}{r} 21 \\ 4\overline{)84} \end{array} \quad \text{means} \quad \begin{array}{r} 2(10) + 1 \\ 4\overline{)8(10) + 4} \end{array}$$

$$\begin{array}{r} 272 \\ 14\overline{)3808} \\ 28 \\ \hline 100 \\ 98 \\ \hline 28 \\ 28 \\ \hline \end{array} \quad \text{means}$$

$$\begin{array}{r} 2(10)^2 + 7(10) + 2 \\ 1(10) + 4\overline{)3(10)^3 + 8(10)^2 + 0(10) + 8} \\ 2(10)^3 + 8(10)^2 \\ \hline 1(10)^3 + 0(10)^2 + 0(10) \\ 9(10)^2 + 8(10) \\ \hline 2(10) + 8 \\ 2(10) + 8 \\ \hline \end{array}$$

The following are found to be essential in the description of the decimal numeration system:

1. Number symbols: 1, 2, 3, 4, 5, 6, 7, 8, and 9.
2. The symbol for zero, 0.
3. A symbol for the base, where *PLACE* value is first introduced. When we move left to the second column (or ten's column), we create our symbol for base ten by writing 10.
4. Position values of symbols, or increasing powers of base ten.

Note that when we see the number 10 (i.e., a one followed by a zero) we assume a base ten, but we will learn that in different number systems 10 can be eight, sixteen, or even two. All that 10 means is that we have denoted the base of the number system under consideration.

Exercises 1-1

Perform the following using expanded notation.

Addition

1. $\begin{array}{r} 143 \\ + \ 23 \\ \hline \end{array}$ 2. $\begin{array}{r} 273 \\ +419 \\ \hline \end{array}$ 3. $\begin{array}{r} 672 \\ +981 \\ \hline \end{array}$

Subtraction

4. $\begin{array}{r} 275 \\ - \ 63 \\ \hline \end{array}$ 5. $\begin{array}{r} 281 \\ -102 \\ \hline \end{array}$ 6. $\begin{array}{r} 543 \\ -258 \\ \hline \end{array}$

Multiplication

7. $\begin{array}{r} 58 \\ \times 16 \\ \hline \end{array}$ 8. $\begin{array}{r} 249 \\ \times \ 65 \\ \hline \end{array}$ 9. $\begin{array}{r} 428 \\ \times 101 \\ \hline \end{array}$

Division

10. $13\overline{)169}$ 11. $19\overline{)855}$ 12. $7\overline{)1000}$

——————— 1-2 THE BINARY NUMERATION SYSTEM

Although the decimal system "fits" our fingers and some types of counting devices, such as the cash register and adding machine, it is not the only system devised by man to do his counting.

Nearly 4000 years ago the Chinese discovered, then forgot, the binary system, or system based on two. The Eskimos counted in a system based on five. The Mayan Indians of Central America also discovered zero and used it in a number system based on twenty. The ancient Babylonians used a numeration system based on sixty.

Leibnitz, a seventeenth century mathematician, is credited with perfecting the binary system as we understand it today. The binary system contains two symbols, 1 and 0, and these symbols fit the "on-off" conditions of an electrical circuit as well as the ten symbols in the decimal system fit our ten fingers.

Because an electric circuit can only be on or off, we can let *one* represent "on" and *zero* represent "off."

If we read the row of lights in Fig. 1-7 from right to left as a binary number, we can describe this number in base two as follows:

Lights	*Binary*	*Decimal*
on	$1 \times 10^0 = 1$	$1 \times 2^0 = 1$
off	$0 \times 10^1 = +0$	$0 \times 2^1 = +0$
on	$1 \times 10^2 = +100$	$1 \times 2^2 = +4$
off	$0 \times 10^3 = +0$	$0 \times 2^3 = +0$
	$\overline{101_{(B)}}$	$\overline{5_{(D)}}$

FIGURE 1-7

The subscript $_{(B)}$ refers to a binary number.

The subscript $_{(D)}$ refers to a decimal number.

Note that in the decimal system we used the symbol 2, but in the binary system we used 10 since there is no digit 2 in the binary system.

In the binary system, each time we move one place to the left, we multiply by two, or double the *place* value of a binary digit.

Binary 1111 means:

Binary		Decimal
1	$1 \times 2^0 =$	1
10	$1 \times 2^1 =$	2
100	$1 \times 2^2 =$	4
1000	$1 \times 2^3 =$	8
1111		15

The binary system fits our description of a numeration system because it has:

1. The number symbol, 1.
2. A symbol for zero, 0.
3. A symbol 10 for the base when we move to the second column to the left.
4. Position values of the symbols as the powers of the base increase.

Note that the largest number before a "carry" in decimal is 9 and the largest number before a "carry" in binary is 1.

If $\qquad\qquad 9 + 1 = 10_{(D)}$

and $\qquad\qquad 1 + 1 = 10_{(B)}$

then $\qquad\qquad 10_{(D)} =$ ten in the decimal system

$\qquad\qquad 10_{(B)} =$ two in the binary system

By comparing our descriptions of the decimal system and the binary system, we can derive a general description of a numeration system N of base n.

Let n be the base of the numeration system. Then N has:

1. number symbols 1, 2, 3, ...$(n - 1)$
2. a symbol for zero, 0
3. a symbol 10 for the base when we move to the second column to the left
4. position values of the symbols as the powers of the base increase.

Using this general definition, we can derive the binary system definitions by substituting two for n.

1-3 CODES THAT RELATE BINARY TO DECIMAL

We noted at the beginning of this chapter that man-computer communications required symbols to exchange information. These symbols are found in codes that have their basis in the binary numeration system.

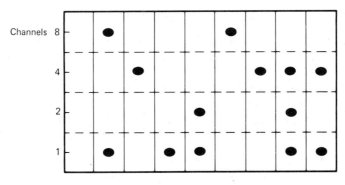

FIGURE 1-8

1-3.1 Binary-Coded Decimal

BCD is a code used by some computers to represent data and per-form calculations. BCD stands for Binary-Coded Decimal. Figure 1-8 shows a punched paper tape containing nine columns and four channels, or rows.

Each punch, or absence of a punch, represents binary data. Using only one column for each digit, combinations of punched holes in channels 8, 4, 2, and 1 can represent any decimal digit 0-9. Either no punches or an 8-punch and a 2-punch could be used to represent zero. In this illustration we have used no punches to represent zero.

If we use the following table,

Table 1-1

Decimal	Binary
0	0000
1	0001
2	0010
3	0011
4	0100
5	0101
6	0110
7	0111
8	1000
9	1001

we can decode the 9-digit number found in Fig. 1-8:

0000	in	column	1	is	0
1001	"	"	2	"	9
0100	"	"	3	"	4
0001	"	"	4	"	1
0011	"	"	5	"	3
1000	"	"	6	"	8
0100	"	"	7	"	4
0111	"	"	8	"	7
0101	"	"	9	"	5

094138475 could be a person's social security number (without the editing dashes) represented in the BCD code.

1-3.2 Excess-Three Code

Another code used by some computers that represents decimal numbers in·a binary format is the Excess-Three Code. In this code the decimal number represented is always its binary equivalent plus 3.

Table 1-2 shows the relationship of the decimal numbers to the excess three code.

Table 1-2

Decimal	Excess-Three
0	0011
1	0100
2	0101
3	0110
4	0111
5	1000
6	1001
7	1010
8	1011
9	1100

1-3.3 Biquinary Code

The Biquinary Code represents any decimal digit with two, and only two, binary digits, or *bits*. This provides a useful method for the computer to verify the validity of binary numbers. If any number has more or fewer than two bits in the "on" status, it will be recognized as an invalid number. For example, 01 0101 would be rejected as an invalid number in this system, since there are three bits in the "on" state.

Table 1-3

Decimal	Biquinary
	50 43210 ← decimal values
0	01 00001
1	01 00010
2	01 00100
3	01 01000
4	01 10000
5	10 00001
6	10 00010
7	10 00100
8	10 01000
9	10 10000

In the *bi*, or two bit column, decimal digits 0–4 require the high-order bit in the "off" or zero status, and decimal digits 5-9 require the low-order bit in the "off" or 0 status. Notice that unlike the BCD and Excess-Three Code, there is no direct relation between code numbers in the Biquinary Code and the binary numeration system. Indirectly, they of course are related, since they both require only the digits 1 and zero.

In the Quinary column for numbers 0–4 and 5–9, a single binary one advances successively one column to the left.

1-3.4 Two-Out-of-Five Code

Another code that uses only two binary characters, or bits, to represent a decimal number from 0–9 is the Two-Out-Of-Five Code. Decimal values for each bit were assigned, and two bits must always be in the "on" state to represent a valid number.

Table 1-4

Decimal	Two-out-of-five
	decimal values → 63210
0	00110
1	00011
2	00101
3	01001
4	01010
5	01100
6	10001
7	10010
8	10100
9	11000

Decimal 4 is represented by a 3-bit and a 1-bit in the "on" state. Decimal 0 could be regarded as a 3 in this scheme, and the selection of 00110 for decimal zero may seem to be out of the designated pattern. However, since we are using a 3-bit and a 0-bit for the number 3, there is no overlapping; and remember that we need two binary ones to express any decimal digit.

Exercises 1-2

1. Using Fig. 1-8 as a guide, write your social security number in BCD.
2. Write the following decimal numbers in Excess-Three Code.

   a. 7   b. 8   c. 0

3. Write the following decimal numbers in the Biquinary and in the Two-Out-Of-Five-Code.

   a. 2   b. 0   c. 7

We have discussed the binary numeration system and four computer codes that relate this numeration system to the decimal system. There are two more important numeration systems used by computers to represent data and to perform calculations. These are the Octal and Hexadecimal systems of numeration, which provide a compact method for handling binary information.

_____ 1-4 THE OCTAL NUMERATION SYSTEM

The octal numeration system is a system based on eight. The octal number $375_{(O)}$ can be described in terms of the powers of base eight as:

Octal		Decimal
5	$5 \times 8^0 =$	5
+ 70	$7 \times 8^1 =$	+ 56
+ 300	$3 \times 8^2 =$	+ 192
$375_{(O)}$		$253_{(D)}$

The subscript $_{(O)}$ refers to an octal number. Because the octal system is based on eight, 8 is *never* used as a symbol in octal notation. In the octal numeration system, eight is represented as 10. Each time we move one place to the left in the octal system, we multiply the place value of the octal digit by eight.

Table 1-5 shows a relationship between the decimal, binary, and octal numeration systems. The octal system fits the description of a numeration system because it has

Table 1-5

Decimal	Binary	Octal
0	000	0
1	001	1
2	010	2
3	011	3
4	100	4
5	101	5
6	110	6
7	111	7
8	1000	10
9	1001	11
10	1010	12
11	1011	13
12	1100	14
13	1101	15
14	1110	16
15	1111	17

1. seven number symbols: 1, 2, 3, 4, 5, 6 and 7.
2. a symbol for zero, 0.
3. the symbol 10 for the base when we move left into the second column.
4. position values for the symbols as the powers of the base increase.

The octal system provides a convenient method of grouping long strings of binary numbers which frequently appear as groups of lights on computer consoles. These lights are the binary representation of some number which may be located internally in the computer. Converting this long string of numbers into a more compact system such as the octal system makes the job of the computer operator, or programmer, easier. Reading out these binary numbers is important both to an operator who wants to determine the next step he should take and to a programmer trying to debug the program.

Binary 10101110111 can be easily converted to an octal number by grouping the digits into groups of three, beginning at the right.

$$\begin{array}{cccc} 010 & 101 & 110 & 111 \quad \text{Binary} \\ 2 & 5 & 6 & 7 \quad \text{Octal} \end{array}$$

To provide a full triplet, a zero has been annexed to the high-order position. The octal number equal to the binary number in each triplet has been written below its corresponding triplet. The highest number we can count up to in base eight, before a "carry" occurs, is 7.

$$6 + 1 = 7$$

and

$$7 + 1 = 10$$

The highest number possible in a binary triplet is 7, represented by three ones, 111.

Binary	010	101	110	111
Octal	2	5	6	7
means	$(2 \times 8^3) + (5 \times 8^2) + (6 \times 8^1) + (7 \times 8^0)$			

Using binary triplets, we can convert at sight any octal number to its binary equivalent by writing a binary triplet for each octal digit. Given $541_{(O)}$, we can write: 5 4 1 in octal means: 101 100 001 in binary. There is no convenient way to group binary digits into the decimal system. If we group the binary digits in fours, we have the hexadecimal system, which we will discuss in the next section.

Exercises 1-3

Convert these binary numbers to octal numbers.

1.   110110011
2.   1010101
3.   1110001101
4.   1010
5.   11010
6.   100000001001
7.   1010110111111
8.   10111011
9.   1111000111011

Convert these octal numbers to binary numbers.

10.   476
11.   1045
12.   201
13.   3321
14.   1001
15.   456
16.   1327

EBCDIC, or the Extended Binary Coded Decimal Interchange Code is used in many of the most recent computers. Its characteristics are therefore very *important* to a programmer. EBCDIC is based upon the relationship of the binary system to the hexadecimal numeration system. To understand this relationship, we will discuss the hexadecimal numeration system.

1-5   THE HEXADECIMAL NUMERATION SYSTEM

The hexadecimal numeration system is a system based on sixteen. The hexadecimal number $134_{(H)}$ can be described in terms of the powers of the base sixteen as:

Hexadecimal		Decimal
4	$4 \times 16^0 =$	4
+ 30	$+ 3 \times 16^1 =$	48
+ 100	$+ 1 \times 16^2 =$	256
$134_{(H)}$		$308_{(D)}$

The subscript $_{(H)}$ refers to a hexadecimal number. Because the hexadecimal system is based on sixteen, 16 is never used as a symbol for the base in hexadecimal notation. Sixteen is represented as 10 in the hexadecimal numeration system. Also, in the hexadecimal system, each time we move one place to the left, we multiply by sixteen, the place value of the hexadecimal digit.

Hexadecimal 1111 means:

Hexadecimal		Decimal
1	$1 \times 16^0 =$	1
10	$1 \times 16^1 =$	16
100	$1 \times 16^2 =$	256
1000	$1 \times 16^3 =$	4096
$1111_{(H)}$		$4369_{(D)}$

We need single digits to represent numbers from ten to fifteen in the hexadecimal system. It has been conventional to use the letters A through F for this purpose. Table 1-6 shows a relationship between the decimal, binary, octal, and hexadecimal systems.

Table 1-6

Decimal	Octal	Binary	Hexadecimal
0	0	0000	0
1	1	0001	1
2	2	0010	2
3	3	0011	3
4	4	0100	4
5	5	0101	5
6	6	0110	6
7	7	0111	7
8	10	1000	8
9	11	1001	9
10	12	1010	A
11	13	1011	B
12	14	1100	C
13	15	1101	D
14	16	1110	E
15	17	1111	F

Although we are using letters A-F, they are *not* used as alphabetic characters. In the hexadecimal numeration system they are considered to be digits representing the decimal numbers from ten to fifteen, A reference to an abacus should provide justification for the use of a single symbol for these quantities.

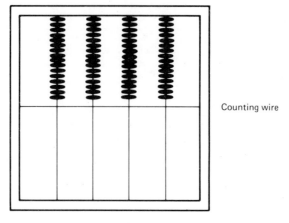

Counting wire

FIGURE 1-9

Figure 1-9 shows a hexadecimal abacus with sixteen beads above the counting wire. Fifteen of these beads must have a unique symbol denoting a digit in the hexadecimal numeration system. When the sixteenth bead is pulled below the counting wire, all of the beads in that column must be pushed above the counting wire and exchanged for one bead in the next column to the left. This creates the symbol for sixteen in the hexadecimal numeration system, 10.

Remember:

$$10 = \text{ten in decimal}$$
$$10 = \text{two in binary}$$
$$10 = \text{eight in octal}$$
$$10 = \text{sixteen in hexadecimal}$$

The hexadecimal system fits our description of a numeration system because it has:

1. the fifteen number symbols: 1, 2, 3, 4, 5, 6, 7, 8, 9, A, B, C, D, E, and F.
2. a symbol for zero, 0
3. the symbol 10 for the base when we move left into the second column.
4. position values for the symbols as the powers of the base increase.

An extension of Table 1-6 shows that counting in hexadecimal follows a pattern similar to the patterns found in the decimal, binary, and octal systems.

Decimal	Octal	Binary	Hexadecimal
16	20	10000	10
17	21	10001	11
18	22	10010	12
19	23	10011	13
20	24	10100	14
21	25	10101	15
22	26	10110	16
23	27	10111	17

The hexadecimal system, like the octal system, provides a convenient method for grouping long strings of binary numbers to make their interpretation "easy on the eyes."

Data stored internally in some computers may be brought out onto "strings" of lights on a console and grouped into quadruplets (groups of four lights). These lights may indicate the status of the program, the presence of data, or an action to be taken by a computer operator. They also act as a debugging tool for programmers when the computer detects an error condition.

The binary number 10101101111111 can be converted to a hexadecimal number by grouping the digits into groups of four, beginning at the right, as follows:

$$0010 \ 1011 \ 0111 \ 1111$$

Notice that two zeros have been annexed to the high-order position to provide a full quadruplet.

Binary	0010	1011	0111	1111
Hexadecimal	2	B	7	F
Decimal	$(2 \times 16^3) + (11 \times 16^2) + (7 \times 16^1) + (15 \times 16^0)$			

The highest number in base sixteen before a "carry" is required is F, and the highest number possible in a binary quadruplet is fifteen.

Using a binary quadruplet, we can convert at sight any hexadecimal number to its binary equivalent. For example, in $FEED_{(H)}$ it is a simple matter to create a binary quadruplet for each hexadecimal digit thus:

F   E   E   D   in hexadecimal

means 1111 1110 1110 1101   in binary

$$FEED_{(H)} = 1111111011101101_{(B)}$$

The subdivision of long strings of binary numbers into quadruplets or triplets is usually indicated on the panels of computer consoles, thus saving computer operators and programmers the inconvenience of copying down these digits and grouping them with commas.

Exercises 1-4

Convert these binary numbers to hexadecimal numbers, and also express them in expanded notation.

1. 10110110
2. 1111000000110
3. 011101
4. 1011101
5. 10111011101
6. 1111111111
7. 1010101010
8. 1001001111011

9. 1011111010101
10. 1101111011101101

Convert these hexadecimal numbers to binary numbers.

11. A12
12. CAB
13. 4F03
14. FED
15. DE161
16. 40F0B

17. Extend Table 1-6 up to decimal 35

1-6 CONVERTING NUMBERS FROM
———— ONE NUMERATION SYSTEM TO ANOTHER

Converting a decimal number to a number in base two, base eight, or base sixteen can be done using repeated division.

The *first dividend* is the decimal number to be converted.

The *divisor* is the base of the new numeration system, expressed as a decimal number.

The *remainder* is a digit in the new number.

1-6.1 Decimal to Binary

Convert $41_{(D)}$ to a binary number, following the explanation just given.

remainder in binary

```
2 )41        1
2 )20        0
2 )10        0
2 ) 5        1
2 ) 2        0
2 ) 1        1
     0
```

Each new quotient becomes the next successive dividend. Division ends with the last dividend greater than zero.

The digits in the remainder form a binary number when they are dropped to the right.

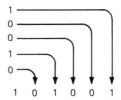

$41_{(D)}$ is 101001 in binary.

1-6.2 Decimal to Octal

Convert $163_{(D)}$ to an octal number

```
8 )163   3
8  )20   4
8   )2   2  4  3
     0
```

$163_{(D)}$ is 243 in octal.

1-6.3 Decimal to Hexadecimal

Convert $285_{(D)}$ to a hexadecimal number.

```
16 )285   D
16  )17   1
16   )1   1  1  D
      0
```

$285_{(D)}$ is 11D in hexadecimal.

1-6.4 Octal to Decimal

The following rules help to convert octal numbers to decimal numbers:
1. Multiply the high-order octal digit by 8.
2. Add the next digit to the right as a decimal number, and express this sum as a decimal number.
3. Multiply this sum by 8.
4. Repeat this procedure until the units digit of the octal number has been added in, but do not multiply this last sum by 8.

Convert $324_{(O)}$ to a decimal number.

$$
\begin{array}{r}
3 \\
\times\,8 \\
\hline
2\,4 \\
+\,2 \\
\hline
2\,6 \\
\times\,8 \\
\hline
2\,0\,8 \\
+\ \ 4 \\
\hline
2\,1\,2
\end{array}
$$

Checking this conversion by expanded notation, $324_{(O)}$ means

$$
\begin{array}{rcl}
4 \times 8^0 & = & 4 \\
+\,2 \times 8^1 & = & 1\,6 \\
+\,3 \times 8^2 & = & 1\,9\,2 \\
\hline
& & 2\,1\,2
\end{array}
$$

in decimal, and the result is verified.

1-6.5 Hexadecimal to Decimal

Following the rules for octal to decimal conversion, substituting 16 for a multiplier, and adding in hexadecimal digits as decimal numbers with decimal sums, convert $4B1_{(H)}$ to a decimal number

$$
\begin{array}{r}
4 \\
\times\,1\,6 \\
\hline
6\,4 \\
+\,1\,1 \\
\hline
7\,5 \\
\times\,1\,6 \\
\hline
1\,2\,0\,0 \\
+\ \ 1 \\
\hline
1\,2\,0\,1
\end{array}
$$

Because $4B1_{(H)}$ means

$$
\begin{array}{rcl}
1 \times 1\,6^0 & = & 1 \\
1\,1 \times 1\,6^1 & = & 1\,7\,6 \\
4 \times 1\,6^2 & = & 1\,0\,2\,4 \\
\hline
& & 1\,2\,0\,1
\end{array}
$$

in decimal, the result is verified.

1-6.6 Binary to Decimal

Binary numbers may be converted to decimal numbers by first converting the binary number to its octal or hexadecimal equivalent and then converting the new octal or hexadecimal number to a decimal number. For example, $1011101_{(B)} = 135_{(O)} = 93_{(D)}$. However, binary numbers may be converted directly to decimal.

Table 1-7

	Powers			
Base	0	1	2	3
Binary	1	2	4	8
Octal	1	8	64	512
Decimal	1	10	100	1000
Hexadecimal	1	16	256	4096

Table 1-7 shows a relationship of the powers of the bases of the four numeration systems under consideration in this chapter. Table 1-7 also may be used as a tool to convert binary, octal, and hexadecimal numbers to decimal numbers. For example, the octal number $435_{(O)}$ can be converted to its decimal equivalent in the following manner:

$$
\begin{array}{rcl}
5 \times 1 & = & 5 \\
3 \times 8 & = & 24 \\
4 \times 64 & = & \underline{256} \\
& & 285
\end{array}
$$

Using Table 1-7, octal 435 is equal to decimal 285. Table 1-7 may be expanded by the student to include higher powers of the bases.

1-6.7 Octal and Hexadecimal Conversions

Using the binary numeration system, we are easily able to convert octal numbers to hexadecimal numbers or hexadecimal numbers to octal numbers.

Octal	4	7	3	0
binary triplets	100	111	011	000

regrouped as

binary quadruplets	1001	1101	1000
Hexadecimal	9	D	8

Thus, $\qquad 4730_{(O)} = 9D8_{(H)}$

As an exercise, convert $FAD_{(H)}$ to its octal equivalent by first writing the correct binary quadruplets, and then regrouping them into binary triplets.

The grouping of binary digits into triplets or quadruplets is important and basic to the internal organization of information in computers. The smallest piece of information stored is usually referred to as a binary digit, or bit. Computers are built with a very large number of bits that hold instructions (in the form of programs) and data that must be stored and processed. If these instructions and data are to be useful, the bits must be easily addressed so that they can be moved, added to, subtracted from, and so on.

To give each bit its own unique address in computer storage would require costly circuitry. Therefore, bits are grouped into units called bytes. A *byte* is the smallest addressable unit of information, and each byte has its own unique internal storage address.

Typically, bytes may contain six or eight bits, depending upon the computer manufacturer's choice of design:

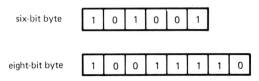

The six-bit byte contains the octal number 51. The eight bit byte contains the hexadecimal number 9E.

The individual bits of a byte are linked electronically so that any unique memory address references the entire group of bits in the byte.

Exercises 1-5

1. Convert the following decimal numbers to numbers in the binary, octal, and hexadecimal systems.
   - a. 25
   - b. 13
   - c. 81
   - d. 44
   - e. 121
   - f. 497
   - g. 5862

2. Convert the following binary numbers to octal, hexadecimal, and decimal numbers.
   - a. 1011011
   - b. 1011001101
   - c. 1011011011011
   - d. 100000111100101
   - e. 10101011011101111
   - f. 1011111111110010111010

3. Convert the following octal numbers to binary, hexadecimal, and decimal numbers.
   - a. 127
   - b. 4610

    c.   55
    d. 103
    e. 4200
    f. 1111

4. Convert the following hexadecimal numbers to binary, octal, and decimal numbers.

    a. DEED
    b.   401
    c.   EA2
    d. BEAD
    e. 129CF

5. Complete the following table.

Decimal	Binary	Octal	Hexadecimal
242			
	1101110		
		374	
			1ED4

FIGURE 1-10

6. Express the following decimal numbers as binary numbers found in bytes.

    a.   362   in the 6-bit format
    b.   432   in the 6-bit format
    c.   904   in the 8-bit format
    d. 1072   in the 8-bit format

2

ARITHMETIC OPERATIONS

We must develop arithmetic skills in the binary, octal, and hexadecimal numeration systems to understand the contents of the registers, arithmetic overflow, signed numbers, and storage maps in computers.

We are concerned in this chapter with the arithmetic operations of addition, subtraction, multiplication, and division.

2-1 BINARY ADDITION

If we were to perform addition on a binary abacus, each wire would contain two beads. Whenever we pulled down two beads in any column, we would force a "carry" into the next column to the left.

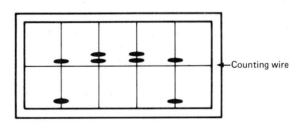

FIGURE 2-1

In Fig. 2-1 we have shown the number nine as $1001_{(B)}$. If we wish to add one to nine, we must note that adding the one in the units column means exchanging two ones for a one in the next column to the left.

$$1001$$
$$+\quad 1$$
$$\overline{1010}$$

Describing this answer in terms of the powers of base two, we have:

Binary		Decimal
0	$0 \times 2^0 =$	0
10	$+ 1 \times 2^1 =$	2
0	$+ 0 \times 2^2 =$	0
1000	$+ 1 \times 2^3 =$	8
$\overline{1010_{(B)}}$		$\overline{10_{(D)}}$

Decimal equivalents are shown to the right of the next two addition examples to make the problems easier to follow.

Binary	Decimal
1010	10
$+\quad 1$	$\underline{1}$
$\overline{1011_{(B)}}$	$\overline{11_{(D)}}$
1011	11
$+\ 100$	$+\ 4$
$\overline{1111_{(B)}}$	$\overline{15_{(D)}}$

The following contain examples of "carry" in binary:

EXAMPLE A   carry digit → 1
$$10101$$
$$\underline{100}$$
$$11001$$

This example may be worked out as $1 + 0 = 1$, $0 + 0 = 0$, $1 + 1 = 10$. Carry the 1 into the next column to the left, where $1 + 0 = 1$. Finally, bring down the last 1 in the fifth column to the left.

EXAMPLE B   carry digits → 11
$$1101$$
$$\underline{100}$$
$$10001$$

This example may be worked out as $1 + 0 = 1$, $0 + 0 = 0$, $1 + 1 = 10$. Carry the 1 into the next column to the left, $1 + 1$ (from the carry) $= 10$. Carry the 1 into the next column to the left.

The next example illustrates a "carry" into a column where two 1's already exist.

EXAMPLE C

```
carry digits → 111
               1011
                111
              10010
```

The operation in the two's column may be expressed as 1 + 1 + 1 = 11 or 1 + 1 + 1 = 1, with a carry of 1. (Observe that in the first column 1 + 1 = 0, with a carry of 1.)

You have frequently added columns of decimal numbers and mentally kept a record of the necessary carries. Although this is not a common situation in binary arithmetic, you should practice this operation to improve your skill.

EXAMPLE D

$$\text{carry digits} \rightarrow \begin{Bmatrix} 111 \\ 1111 \end{Bmatrix}$$

```
              1001
              1111
              0111
              0010
          +   1100
            101101
```

In the method shown above, the carry digits were recorded at the top of each column. In binary this method is awkward because we cannot use any number greater than 1. A faster method is to form answers column by column and then add these answers together.

EXAMPLE E

```
      1001
      1111
      0111
      0010
    +1100
        11    sum of one's column
        11    sum of two's column
        11    sum of four's column
        11    sum of eight's column
    101101
```

We could also perform binary addition in the same way we performed decimal addition on page 25. (Remember that because we are using the binary system of numeration, (10) now means *two*.)

Thus:

		in expanded notation
1011	means	$1(10)^3 + 0(10)^2 + 1(10)^1 + 1$
+ 100	+	$1(10)^2 + 0(10)^1 + 0$
$1111_{(B)}$		$1(10)^3 + 1(10)^2 + 1(10)^1 + 1$

$$
\begin{array}{r}
1001 \\
+\ \ 101 \\
\hline
1110_{(B)}
\end{array}
\qquad \text{means}
$$

$$
\begin{array}{r}
1(10)^3 + 0(10)^2 + 0(10)^1 + 1 \\
+\ \ \ \ \ \ \ \ \ \ \ \ \ \ \ 1(10)^2 + 0(10)^1 + 1 \\
\hline
1(10)^3 + 1(10)^2 + 0(10)^1 + 10 \\
= 1(10)^3 + 1(10)^2 + 1(10)^1 + 0
\end{array}
$$

Exercises 2-1

Do the following binary addition problems.

1. $\begin{array}{r} 1011 \\ +\ \ 101 \\ \hline \end{array}$ 2. $\begin{array}{r} 11101 \\ +\ \ 1001 \\ \hline \end{array}$ 3. $\begin{array}{r} 10011 \\ +\ \ 1111 \\ \hline \end{array}$ 4. $\begin{array}{r} 110111 \\ +110111 \\ \hline \end{array}$ 5. $\begin{array}{r} 1010101 \\ +\ \ 101011 \\ \hline \end{array}$

Add the following columns of binary numbers by either of the two methods discussed.

6. 1011
   0110
   1110
   0101

7. 01100
   10011
   11011
   11100
   10001

Perform the following additions using expanded notation.

8. $\begin{array}{r} 1101 \\ +1001 \\ \hline \end{array}$ 9. $\begin{array}{r} 110110 \\ +\ 10101 \\ \hline \end{array}$ 10. $\begin{array}{r} 11011 \\ +\ 1011 \\ \hline \end{array}$

11. Consider each of the following binary numbers found in a byte, and perform the necessary addition.

a. $\begin{array}{r} 10110111 \\ +00011110 \\ \hline \text{result} \end{array}$ b. $\begin{array}{r} 10101011 \\ 10101011 \\ \hline \text{result} \end{array}$

If you did example 11b correctly, your result should have "overflowed" the allotted 8 bits in the resulting byte. An overflow condition like this may cause a computer program to halt.

Programmers must take care to insure that arithmetic result fields are long enough to accommodate any possible answer. Some coding techniques for overflow handling are given in Chapter 13.

2-2 OCTAL AND HEXADECIMAL ADDITION

Our format for addition provides a similar method for the addition of pairs of numbers in the octal and hexadecimal systems using expanded notation.

In octal: *in expanded notation where* 10 *means eight*

$$
\begin{array}{r}
401 \\
+\ \ 76 \\
\hline
477_{(O)}
\end{array}
\qquad \text{means} \qquad
\begin{array}{r}
4(10)^2 + 0(10)^1 + 1 \\
+\ \ \qquad\quad 7(10)^1 + 6 \\
\hline
4(10)^2 + 7(10)^1 + 7
\end{array}
$$

The next example contains a carry.

$$
\begin{array}{r}
132 \\
+\ \ 55 \\
\hline
207_{(O)}
\end{array}
\qquad \text{means} \qquad
\begin{array}{r}
1(10)^2 +\ \ 3(10)^1 + 2 \\
+\ \qquad\quad 5(10)^1 + 5 \\
\hline
1(10)^2 + 10(10)^1 + 7 \\
= 2(10)^2 +\ \ 0(10)^1 + 7
\end{array}
$$

In hexadecimal: *in expanded notation where* 10 *means sixteen*

$$
\begin{array}{r}
130C \\
+\ \ 491 \\
\hline
179D_{(H)}
\end{array}
\qquad \text{means} \qquad
\begin{array}{r}
1(10)^3 + 3(10)^2 + 0(10)^1 + C \\
+\ \qquad\qquad 4(10)^2 + 9(10)^1 + 1 \\
\hline
1(10)^3 + 7(10)^2 + 9(10)^1 + D
\end{array}
$$

The next example contains a carry.

$$
\begin{array}{r}
A04B \\
+\ \ DAD \\
\hline
+ADF8_{(H)}
\end{array}
\qquad \text{means} \qquad
\begin{array}{r}
A(10)^3 +\ \ 0(10)^2 + 4(10)^1 + B \\
+\ \qquad\quad + D(10)^2 + A(10)^1 + D \\
\hline
A(10)^3 + D(10)^2 + E(10)^1 + 18 \\
= A(10)^3 + D(10)^2 + F(10)^1 +\ \ 8
\end{array}
$$

2-2.1 Displacement

An important reason for studying hexadecimal and octal addition is the need to understand displacement. Before a program can be run, it must be stored somewhere in main memory. Units of main memory called bytes have their own unique addresses, and each computer instruction and each item of data begins at some address in main memory.

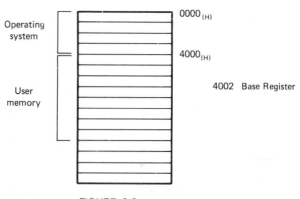

FIGURE 2-2

On many computers the programmer does not know where in main memory his program will be stored at the time it is running. Therefore, he does not know the addresses of all of the bytes his program is using. However, if the address of the first byte in his program were known, then the addresses of other bytes could be calculated.

Figure 2-2 shows a map of main memory and indicates that the first $4000_{(H)}$ positions of main memory are used by the computer's operating system, or the system of programs that monitor and control the operation of the computer. User memory indicates where a user's program may begin. If the operating system requires $4000_{(H)}$ bytes of memory, then, in this example, the user's program may start at the next even-numbered location, or $4002_{(H)}$. This number, $4002_{(H)}$, is placed by the operating system into a register called a base register. The idea is that by giving the programmer an absolute starting address in a register that is accessible to him, the distance between the base address and any other address in a program can be determined. This distance is called a displacement.

Consider $42A0_{(H)}$ as an address found by the programmer in the base register. If he knew that his entire program required $1030_{(H)}$ bytes of storage, then:

$$\begin{array}{r} 42A0 \\ +\,1030 \\ \hline 52D0_{(H)} \end{array}$$

would give the ending address of the program.

As an exercise, consider:

If the programmer knew that the last $200_{(H)}$ bytes in this example were used for data items only, what would be the displacement of the first byte of data?

If these $200_{(H)}$ bytes of data contained eight records of 64 bytes each, what would be the displacement for each of the eight records? Give the answers in hexadecimal.

It is also possible for the programmer to control displacement of individual items of data. Because this subject requires some assembly language programming skills, it is outside the scope of this book.

Exercise 2-2

Figure 2-3 is the decimal addition table.
1. Make an addition table for
   a. the binary system
   b. the octal system
   c. the hexadecimal system

   Include zero in the three tables.

+	1	2	3	4	5	6	7	8	9	10
1	2	3	4	5	6	7	8	9	10	11
2	3	4	5	6	7	8	9	10	11	12
3	4	5	6	7	8	9	10	11	12	13
4	5	6	7	8	9	10	11	12	13	14
5	6	7	8	9	10	11	12	13	14	15
6	7	8	9	10	11	12	13	14	15	16
7	8	9	10	11	12	13	14	15	16	17
8	9	10	11	12	13	14	15	16	17	18
9	10	11	12	13	14	15	16	17	18	19
10	11	12	13	14	15	16	17	18	19	20

FIGURE 2-3

2. Add in octal:

a. 1340   b. 6402   c. 1326   d. 63510   e. 532401
   + 207   + 435   +5507   + 7741   + 65204

f.   316   g. Perform exercises 2a, b, and c in expanded notation.
    427
    553
   +102

3. Add in hexadecimal:

a. 1BEA   b. DEED   c. FEED   d. 4795A
   + 2215   + 3112   + CEED   + B04B

e. 123456   f. 4D
   + ABCDEF   3C
                             50
                            + BB

g. Perform exercises a, b, and c in expanded notation.

4. Describe the addition process, as it would take place in bytes of memory, for exercises 2c and 3a, if they were performed by computers.

2-3 DECIMAL SUBTRACTION

Two methods of subtraction will be presented: 1) the complement method, and 2) the expanded notation method.

2-3.1 The Complement Method

The dictionary defines a complement as something that fills up or completes. For example, for the number 4, 4 + 5 = 9, and we say that 5 is the nine's complement of 4. A ten's complement is a number that, when added to a given number, will yield the complete set for a power of ten. If the given number is 8, then 8 + 2 = 10 and 2 is the ten's complement of 8.

There are Unit Record machines (The accounting machine in Fig. 1-3) still in use that perform decimal subtraction by the complement method.

SAMPLE PROBLEM   Subtract   373   minuend
                               − 185   subtrahend

Take the 9's complement of the subtrahend by subtracting it from a minuend of nines.

$$
\begin{array}{r}
999 \\
-185 \\
\hline
814 \rightarrow \text{nine's complement of 185}
\end{array}
$$

Add in 1 $\dfrac{1}{815}$

815 is the ten's complement of 185, since 815 + 185 yields a complete set for 10^3.

Add the ten's complement of the subtrahend to the original minuend.

$$
\begin{array}{r}
373 \\
+815 \\
\hline
\end{array}
$$

carry out → 1̲]188
of the high
order

The "carry" out of the high order is detected by the circuitry of the accounting machine, and the machine recognizes 188 as a positive number. It is helpful for the student to convert this "carry" out of the high order into a plus sign so that it will not be considered to be a digit in an answer. An algorithm for demonstrating this process follows.

Subtract   373
           − 185

$$373 - 185 = 373 - 185 + (1000 - 1000)$$
$$= 373 - 185 + 999 + 1 - 1000$$
$$= 373 + (\underline{999 - 185 + 1}) - 1000$$

ten's
complement

$$= 373 + 815 - 1000$$
$$= 1188 - 1000 = 188$$

Note that "borrowing" is not necessary when taking the ten's complement.

When may we expect *no* carry out of the high order?

SAMPLE PROBLEM   Subtract   425
                             − 491

We know we will get an answer that will be a negative number.

	999	
Take the nine's complement	− 491	
	508	nine's complement of 491
Add 1	+ 1	
	509	ten's complement of 491
Add in original minuend	+ 425	
No carry out of high order →	934	

The machine detects the absence of a carry out of the high-order position and recognizes the answer as being negative. This also means that the answer itself is in the *complement* form. The number 934 is *not* the result of 425 − 491, but − 66 is. To get the correct answer, − 66, the machine *recomplements* the answer obtained thus far.

	999	
Take the nine's complement	− 934	
	65	
Add 1	+ 1	
	66	ten's complement of 934.

The absence of a carry out of the high order on the first ten's complement procedure indicated that the answer 934 was a negative number and also that it had to be *recomplemented.* After the recomplementation, the final answer can be expressed by the machine as a negative number in its true form, − 66.

SAMPLE PROBLEM   Subtract   4235
                                    − 614

Note that in this problem the subtrahend is one position shorter than the minuend. If we were performing this subtraction in a four-position counter and in a manner similar to the procedures followed by the accounting machine, we would need to pad out the remaining subtrahend position with a zero and add a corresponding amount of 9's in the new minuend.*

$$
\begin{array}{r}
4235 \\
-0614 \\
\hline
9999 \\
-0614 \\
\hline
\end{array}
$$

nine's complement	9385
Add 1	+ 1
	9386
Add original minuend	+ 4235
convert the carry to a + sign	+ 3621

Using the same four-position counter, can we take the ten's complement of 0000?

$$
\begin{array}{r}
9999 \\
- 0000 \\
\hline
9999 \\
+ \quad 1 \\
\hline
10000
\end{array}
$$

The four-position counter still contains our original 0000. Therefore, we say that the quantity zero *cannot* be complemented.

Note that the ten's complement of 9999 is

$$
\begin{array}{r}
9999 \\
-9999 \\
\hline
0000 \\
+ \quad 1 \\
\hline
1
\end{array}
$$
  ten's complement of 9999.

2-3.2 Expanded Notation

Expanded notation can be used in subtraction problems, but it may be necessary at times to regroup or rename the sums of the powers of the

---

* Padding or annexing high-order zeros is done to insure that the complement of the subtrahend has the same number of digits as the minuend.

base. Subtraction using expanded notation will not contain a negative number in the difference in our examples. For example, subtract

$$\begin{array}{r} 126 \\ -117 \\ \hline \end{array}$$

$$
\begin{aligned}
126 &= 1(10)^2 + 2(10)^1 + 6 = 1(10)^2 + 1(10)^1 + 16 \\
-117 &= \underline{1(10)^2 + 1(10)^1 + 7} = \underline{1(10)^2 + 1(10)^1 = 7} \\
& \quad 0(10)^2 + 0(10)^1 + 9 \\
& = 9
\end{aligned}
$$

Exercises 2-3

Perform decimal subtraction by the ten's complement method.

1. 3764	2. 7951	3. 7889	4. 8395	5. 6530
−1045	− 263	−8910	− 26	−6809

Perform the following subtraction problems using expanded notation.

6. 4096	7. 67552	8. 42031
− 185	− 9021	− 1940

2-4 BINARY SUBTRACTION

Complement addition of binary numbers is the method by which most computers perform subtraction. The complement of a binary number is called the *two's* complement. A two's complement (of a binary number) is a number that, when added to a given number, will yield a complete set for a power of two. For example, if the given number is 1, $1 + 1 = 10$. We say that 1 is the two's complement of 1.

To take the two's complement of binary numbers, begin by taking the *one's* complement of the subtrahend, then add 1 to get the two's complement.

SAMPLE PROBLEM   Subtract   1011
$$\underline{-1001}$$

This is performed by subtracting the subtrahend from a minuend of all ones.

$$
\begin{array}{rl}
1111 & \\
\underline{1001} & \\
0110 & \text{one's complement of subtrahend} \\
\underline{+1} & \\
0111 & \text{two's complement of subtrahend}
\end{array}
$$

Another way to perform this procedure is to *invert*, or reverse, each binary digit and then add 1. If the original digit in the subtrahend is a 1, change it to zero; if zero, then change it to 1. Thus, 1001 becomes 0110 + 1 or 0111 without the operation of subtraction. Note that we have padded a zero to the left to give our complemented subtrahend the same number of positions or digits as our minuend.

Now add the two's complement of the subtrahend to the original minuend:

$$
\begin{array}{r}
1011 \\
0111 \\
\hline
\end{array}
$$

carry out of the high order $\Bigg\}$ $+\overline{0010}$
changed to a + sign

If there were no carry out of the high order, our answer would be a negative number and also be in complement form. The computer would detect this condition and recomplement the negative number.

SAMPLE PROBLEM

Subtract   101101

-110001 001110

$+$ 1

001111 ← two's complement of subtra-

$+101101$ ⌐ hend.

no carry out of high order   111100 └original minuend

000011

$+$ 1

-100 recomplemented answer

As an exercise, consider the following binary numbers to be found in bytes and perform the subtraction by the two's complement method.

10111011
-00110111
result

2-5   OCTAL AND HEXADECIMAL SUBTRACTION

The complement addition operations for the octal and hexadecimal numeration systems are the same as they are for the decimal and binary numeration systems.

In the octal system we first take the seven's complement of our subtrahend, then add one.

SAMPLE PROBLEM   Subtract $305_{(O)}$

$- \ \ 27_{(O)}$

take the seven's complement of the subtrahend

$$
\begin{array}{r}
777 \\
\underline{027} \\
750 \\
\underline{+\quad 1} \\
751 \\
\underline{+\,305} \\
+\,256
\end{array}
$$

777
027
750 seven's complement of subtrahend
+   1
751 eight's complement of subtrahend
+ 305 add in original minuend

carry out of → + 256
high order
as + sign

If there were no "carry" out of the high order, the answer would be a negative number in complement form. Of course, this kind of result would have to be recomplemented.

In the hexadecimal system we first take the fifteen's complement of our subtrahend and then add one.

SAMPLE PROBLEM

Subtract $FAD_{(H)}$
          $-\ ADD_{(H)}$

FFF
ADD
522   fifteen's complement of subtrahend
+   1
523   sixteen's complement of subtrahend
+ FAD   add in original minuend

carry out of + 4D0
high order
as + sign

If there were no "carry" out of the high order, the answer would be negative and in complement form. To represent such an answer in true form requires *recomplementing*.

Expanded notation for octal, hexadecimal, and binary subtraction is similar to expanded notation in decimal subtraction.

In octal (where 10 means eight):

$$
\begin{array}{rll}
463_{(O)} = 4(10)^2 + 6(10)^1 + 3 = & 4(10)^2 + 5(10)^1 + 13 \\
-244_{(O)} = 2(10)^2 + 4(10)^1 + 4 = & -2(10)^2 + 4(10)^1 + \ \ 4 \\
\hline
& 2(10)^2 + 1(10)^1 + \ \ 7
\end{array}
$$

In hexadecimal (where 10 means sixteen):

$$
\begin{array}{rll}
12F_{(H)} = & 1(10)^2 + 2(10)^1 + F = & 12(10)^1 + F \\
-\ 4E_{(H)} = - & 4(10)^1 + E = & -\ 4(10)^1 + E \\
\hline
& & E(10)^1 + 1
\end{array}
$$

Exercises 2-4

1. What is the one's complement of 0? of 1?
2. Can you take the ten's complement of 000? Can you take the two's complement of 0?
3. Subtract the following using complement addition.

  In binary:

  a.   10110   b.   1101   c.   10101011
       −10010   −1110   − 0111001

  In octal:

  d.   364   e.   4075   f.   32240
       −211   −4314   − 2051

  In hexadecimal:

  g.   AFC   h.   403B8   i.   DEED
       − 495   − 14A9   − FEED

4. Subtract the following using expanded notation. (Regroup the sums of the powers of the base if necessary.)

  In binary:

  a.   111011   b.   11010   c.   110001
       − 10110   − 1010   − 1111

  In octal:

  d.   375   e.   1064   f.   10432
       − 26   −1023   − 7543

  In hexadecimal:

  g.   13FA6   h.   F7AB0
       −10EAB   −40DCC

5. Express the lowest possible negative number in four digits using its complement form for the following number systems.

  a. decimal
  b. binary
  c. octal
  d. hexadecimal

2-6   MULTIPLICATION

The expanded notation method is often used to perform multiplication in the binary, decimal, octal, and hexadecimal systems. We will illustrate using an example in the octal system (10 means eight).

$$121_{(O)} = \qquad 1(10)^2 + 2(10)^1 + 1$$
$$\underline{\times \ 13_{(O)} =} \qquad \underline{\times 1(10)^1 + 3}$$
$$363 \ = \qquad 3(10)^2 + 6(10)^1 + 3$$
$$\underline{121 \ = 1(10)^3 + 2(10)^2 + 1(10)^1}$$
$$1573_{(O)} = 1(10)^3 + 5(10)^2 + 7(10)^1 + 3$$

Another method of multiplication which can be performed sets up *partial products*. These partial products are then added together to yield the correct answer.

SAMPLE PROBLEM   Multiply in octal   437
$$\times \ 25$$

```
                    437
                  ×  25
                    43        5 × 7 ⎫  Partial products of units
carry ─────────→ ₂ 17         5 × 3 ⎬  digit of multiplier
              → ¹24           5 × 4 ⎭
                    16        2 × 7 ⎫  Partial products of eights
Notice the position of  06    2 × 3 ⎬  digit of multiplier
partial products and the 10   2 × 4 ⎭
zero annexed to a high-  13613
order position as a
placeholder.
```

To develop the student's proficiency in this method, an additional example using partial products in the hexadecimal system is provided.

SAMPLE PROBLEM Multiply in hexadecimal   4A2
$$\times \ B5$$

```
                    4A2
                  ×  B5
                    0A        5 × 2 ⎫  Partial products of units
                    32        5 × A ⎬  digit of multiplier
carry             ¹14         5 × 4 ⎭
                    16        B × 2 ⎫  Partial products of
Notice the position of  6E    B × A ⎬  sixteens digit of
the partial products  ¹2C     B × 4 ⎭  multiplier
                  3468A
```

Binary multiplication is perhaps the easiest of all the multiplication operations discussed in this chapter. Multiplication of binary numbers involves the simple manipulations of shift, copy, and add

$$101101 = 45_{(D)}$$
$$\underline{101} \qquad \underline{5_{(D)}}$$

```
                        copy    101101
          shift one place left  000000
shift one place left and copy   101101
                                11100001 = 225_{(D)}
```

Binary numbers may also be multiplied using expanded notation.

If the student is familiar with all the products of the various pairs of digits in a numeration system, multiplication is as easy in the binary, octal, and hexadecimal systems as it is in the decimal system.

Exercises 2-5

1. Rather than memorize products, the student should construct multiplication tables in

   a. the binary system
   b. the octal system
   c. the hexadecimal system

2. Multiply the following binary numbers using the shift, copy, and add method.

   a. $\quad$ 1011 $\qquad$ b. $\quad$ 101101 $\qquad$ c. $\quad$ 101110
   $\quad\;\times\;$ 111 $\qquad\qquad\;\times\;$ 10001 $\qquad\qquad\;\times\;$ 1010

3. The following binary number is found in a byte.

   $$\boxed{00001010}$$

   Show the result of shifting left 2 positions. This is equivalent to multiplying by _____. If we shifted right one position (padding one high-order zero), the result is equivalent to _____ by _____.

4. Multiply the following octal numbers using the partial products method and expanded notation.

   a. $\quad$ 137 $\qquad$ b. $\quad$ 1234 $\qquad$ c. $\quad$ 707
   $\quad\;\times\;$ 20 $\qquad\qquad\;\times\;$ 56 $\qquad\qquad\;\times\;$ 45

5. Multiply the following hexadecimal numbers using the partial products method and expanded notation.

   a. $\quad$ 1AD $\qquad$ b. $\quad$ F0F0 $\qquad$ c. $\quad$ C128
   $\quad\;\times\;$ 63 $\qquad\qquad\;\times\;$ DAB $\qquad\qquad\;\times\;$ A09

2-7 DIVISION

Division can be regarded as a series of repeated subtractions. Using this method of division in a decimal problem, we get

```
34)8194
   3400      34 × 100
   ────
   4794
   3400      34 × 100
   ────
   1394
    340      34 × 10
   ────
   1054
    340      34 × 10
   ────
    714
    340      34 × 10
   ────
    374
    340      34 × 10
   ────
     34
     34      34 × 1
   ────
```

Adding up the number of times 34 has been subtracted from 8194, we get 100 + 100 + 10 + 10 + 10 + 10 + 1 = 241.

In the hexadecimal system, division by repeated subtractions is difficult, but the hexadecimal multiplication table constructed in Exercises 2-5, 1 should provide sufficient help in working out hexadecimal division problems.

SAMPLE PROBLEM Divide in hexadecimal A)B5E.

```
A)B5E
  A00      A × 100
  ───
  15E
   A0      A × 10
  ───
   BE
   A0      A × 10
  ───
   1E
    A      A × 1
  ───
   14
    A      A × 1
  ───
    A
    A      A × 1
  ───
```

Adding up the number of times A has been subtracted from B5E, we have $100 + 10 + 10 + 1 + 1 + 1 = 123_{(H)}$.

Binary division makes the method of repeated subtractions easier to perform.

SAMPLE PROBLEM   Divide in binary $10\overline{)11110}$

$$
\begin{array}{r}
1111 \\
10\overline{)11110} \\
\underline{10} \\
11 \\
\underline{10} \\
11 \\
\underline{10} \\
10 \\
\underline{10}
\end{array}
$$

Note that division by two is similar to shifting right one position.

A remainder in a binary, octal, or hexadecimal division problem is treated in the same manner as a remainder in decimal division.

$$
\begin{array}{r}
11 \qquad 11 \\
111\overline{)11000} \quad \overline{111} \leftarrow \text{Fraction} \\
\underline{111} \\
1010 \\
\underline{111} \\
11
\end{array}
$$

The expanded notation method is illustrated with a problem in the octal system.

SAMPLE PROBLEM

Divide the octal numbers $12\overline{)4110}$ in expanded notation. Note that 10 is written for base eight.

$$
\begin{array}{r}
3(10)^2 + 2(10)^1 + 4 \\
1(10)^1 + 2\overline{)4(10)^3 + 1(10)^2 + 1(10)^1 + 0} \\
\underline{3(10)^3 + 6(10)^2} \\
3(10)^2 + 1(10)^1 \\
\underline{2(10)^2 + 4(10)^1} \\
5(10)^1 + 0 \\
\underline{5(10)^1 + 0}
\end{array}
$$

If computers perform

multiplication by shifting and adding
division by repeated subtraction
subtraction by complement addition

it should be clear that addition is their basic arithmetic operation.

Exercises 2-6

1. Divide these binary numbers using repeated subtractions.
   a. 11)$\overline{11011}$
   b. 110)$\overline{10100}$
   c. 101)$\overline{1010101}$

2. Divide these octal numbers using both the repeated subtraction method and the expanded notation method.
   a. 12)$\overline{11174}$
   b. 45)$\overline{7200}$
   c. 37)$\overline{25376}$

3. Divide these hexadecimal numbers using both the repeated subtraction method and the expanded notation method.
   a. E)$\overline{C4}$
   b. 2C)$\overline{15088}$

4. Given: 01011000

   Show the result of dividing by 2. How many positions did we shift right? Suppose the hexadecimal number 18 was the quotient (with no remainder) obtained by shifting right 3 times. What was the original dividend?

2-8 SHIFTING BINARY DIGITS

Multiplication and division by shifting binary bits left or right in a register present the following interesting situation.

Consider a register containing the hexadecimal value

> 0CA0

What would the result be of dividing by two, or performing one shift to the right and padding high-order zeros?

To divide by four, we must shift right two binary bits and pad high-order zeros to the left. What is the result of dividing the register above by four, i.e., shifting right two positions?

Give the result for multiplying the register above by eight.

It is easy to see that we can alter hexadecimal values in registers by shifting and thereby alter the coding for instructions or data in a program. This is one method by which programmers can maintain privacy of files. For example, if at any time someone tried to read out the contents of main memory, and the operating system altered the hexadecimal values in a unique manner, no one could be sure of the program's instructions or data.

3

SET THEORY

We noted in the Introduction that a *set* is a collection of elements. These elements can be numbers, persons, objects or anything else. The tires on a car form a set, the days of the week, all the negative numbers, all students in a class, and so on. Typically, a set is a number of things grouped in such manner·as to form a whole.

3-1 PROPERTIES OF ELEMENTS IN A SET

Properties of elements in sets are:

1. There must be a statement which states whether an element is or is not in a set.
2. The order in which the elements appear in a set is not important.
3. Elements in a set are distinct.

Definition 3-1

Two sets are equal if and only if they contain the same, i.e., identical, elements.

This means the same as the following: if two sets are equal, then they contain the same elements; and if they contain the same elements, then the two sets are equal. Thus, if

$$A = \{1, 3, 5\}$$

and

$$B = \{5, 1, 3\}$$

then $A = B$ by Definition 3-1, and by property 2.

Two sets may contain the same *number* of elements, yet not be equal.

3-1.1 One-to-One Correspondence

It is believed that our ancient ancestors made a one-to-one correspondence between objects they were interested in (or a set of things) and the fingers of their hands (another set). Today, the meaning for a one-to-one correspondence is much the same.

For example, let $A = \{1, 2, 3, 4\}$ and $B = \{a, b, c, d\}$; then we can show a one-to-one correspondence between A and B in the following manner:

$$\{1, \ 2, \ 3, \ 4\}$$
$$\updownarrow \quad \updownarrow \quad \updownarrow \quad \updownarrow$$
$$\{a, \ b, \ c, \ d\}$$

This is only one of many one-to-one correspondences possible between A and B.

Definition 3-2

Two sets are *equivalent* if each element of the first set can be paired with one and only one element of the second set and if each element of the second set can be paired with one and only one element of the first set. That is, if there is a one-to-one correspondence between the elements of the two sets, the sets are equivalent.

Equivalent sets are sets that contain the same number of elements. If two sets are equal then they are equivalent. However, the converse is not necessarily true because the definition for equality states that identical elements must exist in both sets.

EXAMPLE $\qquad\qquad A = \{\Delta, \$, ?\}$
$\qquad\qquad\qquad\qquad\quad B = \{red, green, blue\}$

Since the elements in A and B can be put into a one-to-one correspondence, A and B are equivalent sets by Definition 3-2; but $A \neq B$.

EXAMPLE $\qquad\qquad A = \{1, 2, 3\}$
$\qquad\qquad\qquad\qquad\quad B = \{3, 2, 1\}$

In this example, *A* and *B* are equal and equivalent sets by Definitions 3-1 and 3-2, respectively.

A set may contain no elements, a definite number of elements, or an infinite number of elements.

Definition 3-3

A set containing no elements is called the *null* set or empty set and may be written as either $A = \{\ \ \}$ or $A = \emptyset$.

A set containing a definite number of elements can be written in the *listing format*

$$A = \{4,\ 5,\ 6\},$$

or by the *set builder* notation:

$$A = \{x \mid 3 < x < 7,\ x \text{ is a natural number}\}.$$

The notation above reads: set *A* = the set of all *x* such that *x* is a natural number greater than 3 and less than 7.

Definition 3-4

A set is said to contain an infinite number of elements if it can be put into a one-to-one correspondence with:

$$A = \{1,\ 2,\ 3,...\}$$

where *A* is the set of all of the counting, or natural, numbers.

The use of three dots may also indicate omitted elements in a set containing a finite number of elements. In the set $B = \{1, 2, 3,..., 50\}$ the three dots are a short-cut method to indicate the missing natural numbers from 4 through 49.

When discussing sets, it is useful to have some general reference in mind from which all the elements of the sets can be drawn. Consider, for example, the set of all programmers. This consists of Cobol programmers, Fortran programmers, Assembly-Language programmers, and in general, people who simply want to increase their skills in getting computers to help them solve problems. There are also American programmers, French programmers, etc. Therefore, from a general set of all programmers, we can draw any one of a number of different subsets of programmers. A general set is sometimes referred to as a Universal Set.

3-1.2 Universal Set

The term universal set is used to designate the entire collection of things under consideration. To demonstrate this concept in a practical data processing situation, assume that the entire collection under consideration is the set of data items in a pay check. Then

Paycheck = {gross pay, hours worked, rate of pay, medical insurance, savings bonds, social security number, state taxes, federal taxes, FICA, net pay.}

Paycheck is the total collection of items under consideration, but it is clear that there are smaller collections of items within the larger collection. Involuntary deductions or taxes such as state and federal taxes and FICA deductions, and voluntary deductions such as medical insurance and savings bonds make up two smaller collections called *subsets*.

All employee records are contained in a file. A file is a collection of records that have common characteristics. Consider individual employees with payroll records. A record is a collection of related data items. A data item is simply a piece of information residing in a field in the record. A field is one or more contiguous positions in a record.

Table 3-1

	Name	Gross Pay	Hours	Rate	Medical Insurance	State Tax	Fed Tax	Fica	Savings Bonds	Net Pay
Record 1	A	960.00	80	12.00	18.12	40.00	190.01	58.85	92.31	560.71
Record 2	B	1040.00	80	13.00	18.12	42.40	215.61	69.16	100.00	594.71

Table 3-1 represents a simplified data base for employees or the collection of data items needed to prepare payroll checks. In a payroll situation, sets are a useful concept. Observe that the elements of the set, gross pay, are calculated for each individual by multiplying two other sets: hours and rate. Net pay is calculated by subtracting out the deductions for medical insurance, savings bonds, state, federal, and FICA taxes. Further, the subsets comprising taxes must be totalled for all employees and checks have to be sent to the appropriate government agencies.

Using our nonmenclature of upper and lower case letters, let U represent the total collection of paycheck elements, and let g = gross pay, w = hours worked, r = rate of pay, m = medical insurance, b = savings bonds, n = social security number, s = state taxes, f = federal taxes, t = FICA, and p = net pay. Then

$$U = \{g, w, r, m, b, n, s, f, t, p\}$$

The symbol U is often used to represent the Universal Set.

3-1.3 Subsets

If we let T represent a smaller collection of items which make up the involuntary deductions called taxes, then

$$T = \{s, f, t\}$$

where all the elements of T are included in U. However, there are elements in U not found in T.

Definition 3-5

If every element of a given set A is an element of a given set B and at least one element in B is not found in A, then A is said to be a *proper subset* of B. This can be denoted as $A \subset B$.

In the paycheck universe we can write $T \subset U$, or $\{s, f, t\}$ $\subset \{g, w, r, m, b, n, s, f, t, p\}$. The symbol $\subset$ is read "is a proper subset of."

The null set, $\{\ \ \}$ or $\emptyset$, is a *proper* subset of every set except itself. For example:

$$\emptyset \subset T, \text{ also } \emptyset \subseteq \{s, f, t\}$$

Definition 3-6

Set A is said to be a subset of set B if and only if each element of A is an element in B. This is denoted as $A \subseteq B$.

This means that

$$\{s, f, t\} \subseteq \{s, f, t\}$$

where the symbol $\subseteq$ is read "is contained in," or "is a subset of." Therefore, every set is a subset of itself.

Further, $\{s, f, t\} \subseteq \{g, w, r, m, b, n, s, f, t, p\}$ is also true because T is contained in U or $T \subseteq U$. Similarly $U \subseteq U$ and $\emptyset \subseteq \emptyset$.

Exercises 3-1

1. Write each of the following sets using braces and listing the elements or members.

   a. days of the week
   b. natural numbers between 7 and 12
   c. natural numbers less than 3 and greater than 8.
   d. even natural numbers
   e. women quarterbacks in professional football

2. A is the set of all counting numbers less than or equal to 10. State which of the following expressions are true, and which are false.

   a. $8 \in A$
   b. $.5 \in A$
   c. $25 \notin A$
   d. $\emptyset \in A$
   e. $\emptyset \subset A$

   f. $\{\ \ \} \subseteq A$
   g. $12 \in A$
   h. $\frac{1}{4} \notin A$
   i. $0 \subseteq A$
   j. $-1 \in A$

3. Which of the following sets are finite? Which are infinite?

   a. the natural numbers
   b. the population of Missouri
   c. the decimal places for π
   d. all the money in America

4. If $U = \{\$, a, ?\}$, list all of the possible subsets in U.
5. Which pairs of sets are equivalent? Which are equal?

   a. $\{a, b, c,\}$ and $\{c, a, b\}$
   b. $\{1, 2, 3, 4\}$ and $\{5, 6, 7, 8\}$
   c. $\{\frac{1}{2}, \frac{2}{3}, \frac{7}{8}\}$ and $\{x, y, z\}$
   d. $\{R, w, *\}$ and $\{$dogs, cats$\}$

6. Let $U = \{1, 2, 3, 4, 5, 6, 7\}$, $A = \{1, 2, 3\}$, and $B = \{2, 3, 4, 5\}$. Using $\in$, $\notin$, $\subset$, $\subseteq$, $\not\subset$, and $\not\subseteq$, describe the relationship of the following sets. (**Example:** The relationship of A and U is $A \subset U$.)

   a. B, U d. $1, A$
   b. A, B e. $\emptyset, B$
   c. $7, B$

7. Let $A \subset U$, $B \subset U$, $C \subseteq B$, $x \in A$, and $y \in B$.

   a. Can $x \in B$?
   b. Must $y \in A$?
   c. Can $y \in C$?
   d. Is $C \subset U$?

8. List all subsets of each set.

   a. $A = \{1, 2\}$ set of 2 elements
   b. $B = \{1, 2, 3\}$ set of 3 elements
   c. $C = \{1, 2, 3, 4\}$ set of 4 elements

9. Based on your answers to Exercise 8, determine the number of subsets in a set of 5 elements. Can you determine a formula for the number of subsets in a set of n elements?
10. Is there a set which has only one subset?
11. Describe the following sets.

   a. $\{\ \}$
   b. $\emptyset$
   c. $\{\emptyset\}$
   d. $\{\{\emptyset\}, \emptyset\}$

12. List the subsets for a, b, c, d in Exercise 11.

3-2   OPERATIONS ON SETS

Earlier, familiar arithmetic operations on numbers were considered. Operations may also be performed on sets. These operations, called union and intersection, can be graphically described through the use of Venn diagrams. (See Fig. 3-1).

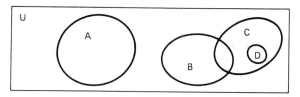

FIGURE 3-1

Although the compartments vary in many possible ways, there is no section which can remain unspecified as to membership. If an element is common to more than one set in the rectangle, it will only appear once in an operation on the sets to which it belongs. (See Fig. 3-2.) If the element is employee deductions, for example, the Venn diagram is only concerned with this logical type, not how many employee deductions there are. In Fig. 3-1 the universe or entire collection of things under discussion is indicated by all points inside the rectangle U.

The different shapes indicate the subsets of U, and in Fig. 3-1 $A \subset U$, $B \subset U, C \subset U, D \subset U$, and $D \subset C$. All shapes A, B, C, and D are "properly contained" in U. D is also properly contained in C. B and C have some commonality, and A has no elements common to B, C, or D.

Definition 3-7

If two sets A and B are not empty sets and contain no common elements, they are said to be *disjoint sets*.

Let $A = \{1, 2\}$ and $B = \{3, 4\}$.
Then $1 \in \{1, 2\}$
and $2 \in \{1, 2\}$;
but $1 \notin \{3, 4\}$
and $2 \notin \{3, 4\}$.
Thus, A and B are disjoint sets because they have no elements in common.

There are, as noted earlier, two operations on the subsets of a given universe that are important to us.

3-2.1 Union

Definition 3-8

The *union* of two subsets, A and B, is a *binary operation* that yields a third subset C formed by the elements found in either A or B, or in both A and B.

The symbol ∪ indicates the union of two subsets, and A ∪ B is read "A union B," or "the union of A and B."

To illustrate the properties of the operation of union, we will use the universal set, S = {1, 2, 3},* with the following proper subsets:

$$A = \{1, 2\}$$
$$B = \{1, 3\}$$
$$C = \{2, 3\}$$
$$D = \{1\}$$
$$E = \{2\}$$
$$F = \{3\}$$
$$\varnothing = \{ \ \}$$

Then

$$A \cup B = \{1, 2, 3\}$$

or

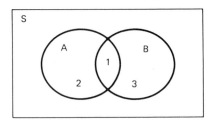

FIGURE 3-2

The elements in A ∪ B are listed only once because two sets have been joined, not added in any numerical sense. This conforms to the third property of sets. Because we are interested only in classes of objects, the repetition of the digit 1 is redundant. In the example we are interested only in the fact that the digit 1 is an element in either A or B or in *both* A and B.

Table 3-2, called a union table, shows the results for all possible unions in the universe S, where S = {1, 2, 3}.

Because the union of any two sets produces a subset of the universal set, the operation of union has the property of *closure*. (That is, the union of any two sets in the universal set yields a set in the universal set or the universal set.) Studying Table 3-2, note that B ∪ C = C ∪ B, F ∪ D = D ∪ F, A ∪ B = B ∪ A, etc., i.e., the Commutative Law holds true for the union of sets. See Definition 3-8.

---

* We use the letter S instead of U to designate the universal set to avoid confusion with the union symbol, ∪.

The Commutative Property

Inspection of the elements in the subsets being operated on also demonstrates the *commutative property* of union.

Table 3-2

∪	S	A	B	C	D	E	F	φ
S	S	S	S	S	S	S	S	S
A	S	A	S	S	A	A	S	A
B	S	S	B	S	B	S	B	B
C	S	S	S	C	S	C	C	C
D	S	A	B	S	D	A	B	D
E	S	A	S	C	A	E	C	E
F	S	S	B	C	B	C	F	F
φ	S	A	B	C	D	E	F	φ

Since $B \cup C = C \cup B$,

$$\{1, 3\} \cup \{2, 3\} = \{1, 2, 3\}$$

and $\qquad \{2, 3\} \cup \{1, 3\} = \{1, 2, 3\}$

Therefore $\qquad \{1, 3\} \cup \{2, 3\} = \{2, 3\} \cup \{1, 3\}$

Since in Table 3-2 the arrangement of sets is symmetrical, the left diagonal (\\), sometimes called the major diagonal, provides a way of locating the sets that demonstrates the commutative law. Thus, if we place a pencil along the diagonal, we can see that the subsets to the right of the pencil form a mirror image of the subsets on the left.

Table 3-2a

∪	S	A	B	C	D	E	F	φ
S	S	S	S	S	S	S	S	S
A	S	A	S	S	A	A	S	A
B	S	S	B	S	B	S	B	B
C	S	S	S	C	S	C	C	C
D	S	A	B	S	D	A	B	D
E	S	A	S	C	A	E	C	E
F	S	S	B	C	B	C	F	F
φ	S	A	B	C	D	E	F	φ

From this inspection, we can see that $B \cup A = A \cup B$, $E \cup B = B \cup E$, and so on.

The Associative Property

Consider $(D \cup E) \cup F = D \cup (E \cup F)$ where D, E, and F are as shown in Table 3-2. As in algebra, the parentheses indicate the operation to be performed first.

Since

$$(D \cup E) = A \text{ and } (E \cup F) = C,$$

it follows that

$$A \cup F = D \cup C.$$

Also, since

$$A \cup F = S \text{ and } D \cup C = S,$$

then

$$(D \cup E) \cup F = D \cup (E \cup F).$$

The Venn diagrams in Fig. 3-3 graphically illustrate this associative property.

From Table 3-2 it can be shown that the associative property holds for the union of all sets in the universe S.

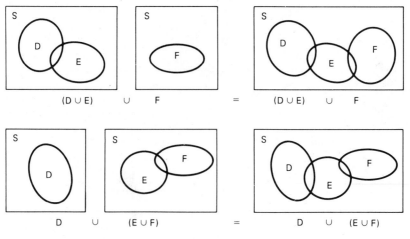

FIGURE 3-3

The Identity Property

The identity property for union can also be derived from Table 3-2, where:

$$S \cup \emptyset = S$$
$$A \cup \emptyset = A$$
$$B \cup \emptyset = B$$
$$C \cup \emptyset = C$$
$$D \cup \emptyset = D$$
$$E \cup \emptyset = E$$
$$F \cup \emptyset = F$$
$$\emptyset \cup \emptyset = \emptyset$$

From this we see that $\emptyset$ serves as the identity for the union operation. See the definition of the identity element on page 5.

Observe from the last column to the right or in the bottom row of Table 3-2, the union of a given set and the null set always yields the given set.

Exercises 3-2

1. Given:

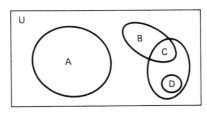

FIGURE 3-4

Name the pairs of disjoint sets.
2. Draw the Venn diagram to illustrate the commutative law for union in $A \cup B = B \cup A$.
3. Find the result for

   a. $\{1, 2, 3\} \cup \{2, 3, 4\}$
   b. $\{x \mid 6 < x \le 9\} \cup \{0, 3, 80\}$ (where x is a natural number)
   c. $\{a, b, c, d\} \cup \{a, c, e\}$
4. Verify the associative property for
   $A = \{\$, \#\}, B = \{11\}, C = \{/, \#\}$
   where $(A \cup B) \cup C = A \cup (B \cup C)$
5. Using Venn diagrams, prove that $\emptyset \cup A = A$, for any set.
6. Given $S = \{\Delta, 0\}$, list the four subsets of S and construct the appropriate union table.

7. When does

    a. $A \cup D = A$

    b. $A \cup \emptyset = U$

8. Each customer of the XYZ Retail Store has two types of records in the store's files: Record 1, which contains the customer's name, address, and customer identification number, and is used for mailing lists when advertising sales; and Record 2, which contains the customer's identification number, purchases for the last month, date and amount of the last payment, and balance due. This record type is used to bill customers.

    Consider each record as a collection of data items.

a. use the listing format to indicate all of the elements formed by the union of these sets (or records).

b. Does the union of these two sets demonstrate the property of closure?

c. Draw the Venn diagram to illustrate the elements, showing any elements common to the two sets.

3-2.2 Intersection

If set $B = \{1, 2, 3, 4\}$ and set $C = \{2, 3, 4, 5\}$, the elements common to both B and C are 2, 3, and 4. If $E = \{2, 3, 4\}$ then set E contains only those elements found in both B and C.

Definition 3-9

The *intersection* of two subsets A and B is a binary operation that yields a third subset C formed by the elements found in both of the subsets A and B.

The symbol $\cap$ indicates the intersection of two subsets, and $A \cap B$ is read "A intersect B," or "the intersection of A and B." For the sets given in the example above, we can write

$$\{1, 2, 3, 4,\} \cap \{2, 3, 4, 5\} = \{2, 3, 4\}, \text{ or } B \cap C = E$$

To illustrate the properties of the operation of intersection, we will use the universal set $S = \{\$, \#, /\}$ with the following proper subsets:

$$A = \{\$, \#\}$$
$$B = \{\$, /\}$$
$$C = \{\#, /\}$$
$$D = \{\$\}$$
$$E = \{\#\}$$
$$F = \{/\}$$
$$\emptyset = \{ \ \}$$

Then

$$A \cap B = \{\$\} \text{ or } A \cap B = D$$

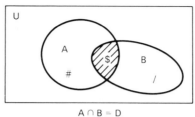

A ∩ B = D

FIGURE 3-5

Table 3-3

∩	S	A	B	C	D	E	F	Ø
S	S	A	B	C	D	E	F	Ø
A	A	A	D	E	D	E	Ø	Ø
B	B	D	B	F	D	Ø	F	Ø
C	C	E	F	C	Ø	E	F	Ø
D	D	D	D	Ø	D	Ø	Ø	Ø
E	E	E	Ø	E	Ø	E	Ø	Ø
F	F	Ø	F	F	Ø	Ø	F	Ø
Ø	Ø	Ø	Ø	Ø	Ø	Ø	Ø	Ø

In the Venn diagram shown in Fig. 3-5, the shaded area indicates the subset resulting from the intersection of A and B.

The results for all possible intersections in the universe S are shown in Table 3-3.

The operation of intersection, like the operation of union, has the property of closure, because the intersection of any two sets produces a subset of the universal set.

The Commutative Property

Inspection of Table 3-3 reveals that $D \cap C = C \cap D$, $B \cap F = F \cap B$, $B \cap E = E \cap B$, etc. Thus, the commutative property holds for the intersection of sets. This follows from the definition of the intersection operation and can be observed by inspection of the elements in the subsets being operated on.

EXAMPLE $E \cap A = A \cap E$:

$$\{\#\} \cap \{\$, \#\} = \{\#\}$$

and

$$\{\$, \#\} \cap \{\#\} = \{\#\}$$

therefore,

$$\{\#\} \cap \{\$, \#\} = \{\$, \#\} \cap \{\#\}$$

In Table 3-3 the arrangement of sets is symmetrical. Therefore, in this table, like in the union Table 3-2, the left diagonal (\) provides a way of locating the sets that demonstrate the commutative law. We find $B \cap A = A \cap B$, $C \cap A = A \cap C$, and so on. Compare Table 3-2, which illustrates the union operation, to Table 3-3, which illustrates the intersection operation. The left diagonal in Table 3-3 is the same as it appeared in Table 3-2, but in Table 3-2 the right diagonal (/), sometimes called the minor diagonal, contains only the universal set, S. This same diagonal in Table 3-3 contains only the null set.

When two subsets have no elements in common, their intersection yields the null set, $\varnothing$. Such sets, as we have seen in Definition 3-7, are commonly called *disjoint*.

The Associative Property

The associative property also holds true for the intersection of sets. Refer to Table 3-3 and consider the example:

$$(C \cap B) \cap F = C \cap (B \cap F)$$

Since

$$C \cap B = F \text{ and } B \cap F = F$$

we have

$$F \cap F = F \text{ and } C \cap F = F$$

Therefore

$$(C \cap B) \cap F = C \cap (B \cap F)$$

The Venn diagrams in Fig. 3-6 show associativity for $(A \cap B) \cap D = A \cap (B \cap D)$.

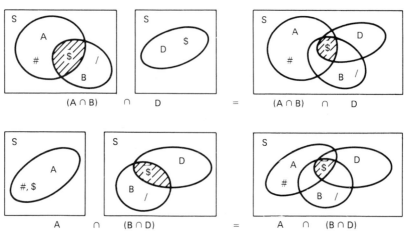

FIGURE 3-6

The Identity Property

The identity property for intersection can be noted in Table 3-3 where

$$A \cap S = A$$
$$B \cap S = B, \text{ etc.}$$

The universal set S serves as the identity for the intersection operation. This can be determined by inspection of the sets in the second row or in the second column from the left, where the intersection of a given set with the universal set always yields the given set.

Note also that the intersection of a given set with itself yields the given set.

EXAMPLE $\qquad A \cap A = A$

As an exercise, the student should demonstrate why sets A, B, C, D, E, and F cannot be called the identity for intersection operations in Table 3-3.

3-3 COMPLEMENT

For every subset in a universal set, there is another subset that contains just those elements that the first one does not.

For example, in $U = \{1, 2, 3, 4\}$, if $A = \{3, 4\}$, then a set B exists such that $B = \{1, 2\}$. This relationship is called the *complement.*

Definition 3-10

The complement of a set A in U is another set containing all of the elements in U not found in A.

The symbol $\overline{A}$ or A' denotes the complement of A. We can illustrate this distinction in Venn diagrams. In Fig. 3-7 the shaded portion is identified as $\overline{A}$ and represents those elements not found in A.

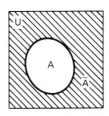

FIGURE 3-7

3-4 DISTRIBUTIVE LAWS

Earlier we learned about a distributive law (page 5) in which multiplication was distributive with respect to addition. Consider

$$x(y + z) = xy + xz$$

If $\quad\quad\quad\quad\quad\quad x = 2, y = 3, z = -1,$

then $\quad\quad\quad\quad 2[3 + (-1)] = 2 \cdot 3 + 2 \cdot (-1)$

or $\quad\quad\quad\quad\quad\quad\quad\quad 4 = 4$

Operations on sets involve two distributive laws.

3-4.1 First Distributive Law

The Venn diagrams in Fig. 3-8 graphically illustrate that intersection is distributive with respect to union. In the algebra of sets we can also state that intersection is distributive with respect to union:

$$X \cap (Y \cup Z) = (X \cap Y) \cup (X \cap Z)$$

This says that intersecting the union of two sets Y and Z by a third set X yields the same set as intersecting sets Y and Z by set X and joining the resulting sets by the union operation.

3-4.2 Second Distributive Law

In conventional algebra we do not have the equality $x + yz = (x + y)(x + z)$. (Addition is not distributive with respect to multiplication.)

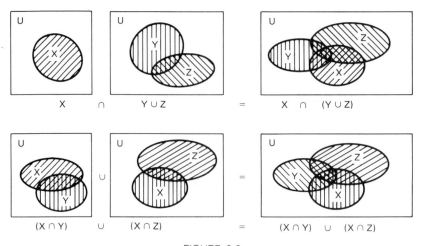

FIGURE 3-8

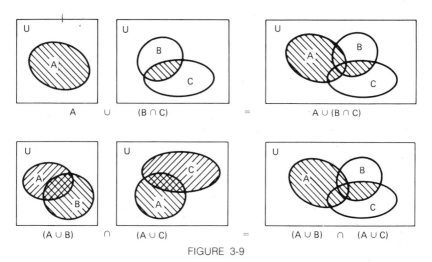

FIGURE 3-9

Substituting as before we can demonstrate that

$$2 + (3 \times -1) \neq (2 + 3)(2 - 1)$$

However, in the algebra of sets, we can state that $A \cup (B \cap C) = (A \cup B) \cap (A \cup C)$, or that union is distributive with respect to intersection. This is illustrated in Fig. 3-9. It is called the second distributive law. It means that joining by union set A to the set formed by the intersection of sets B and C yields the same set as joining by union sets B and C with set A and intersecting the resulting sets. Note that interchanging the union and intersection operations of the first distributive law results in the second distributive law.

Exercises 3-3

1. Form the intersection of the following sets.
   a. $\{1, 2, 4\} \cap \{1, 3, 5, 6\}$
   b. $\{\#, ?, @\} \cap \{\#, ?, !\}$
   c. $\{x|\ 7 \geq x \text{ or } x > 9\} \cap \{1, 2, 3\}$ where x is a natural number
   d. {Even natural numbers less than 21} ∩ {Odd natural numbers less than 10}

2. *Given* $U = \{a, b, c, d, e, f, g, h\}$
   $A = \{b, d, f, h\}$, $B = \{a, b, c, d\}$
   $C = \{a, c, e, g\}$
   List the elements in
   a. $A \cap B$
   b. $A \cup C$
   c. $\overline{A}$
   d. $\overline{B}$

e. $\overline{(A \cup B)}$

f. $\overline{A} \cap C$

g. $\overline{B} \cap \overline{C}$

h. $A \cup (B \cap C)$

Perform the operation inside the parentheses first.

3. Given $S = \{1, 2, 3, 4\}$, list all the subsets and construct the intersection table for all intersection operations on this set.

4. Given

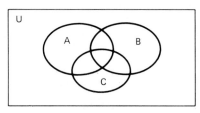

FIGURE 3-10

Shade those parts of the diagram denoted by the following

a. $A \cap B$

b. $A \cap \overline{B}$

c. $\overline{A} \cup B$

d. $\overline{C} \cap \overline{B}$

e. $(A \cup B) \cap \overline{C}$

f. $(A \cap B) \cup (\overline{A} \cap C)$

5. When does

a. $A \cap B = A \cup B$?

b. $A \cap \emptyset = \emptyset$?

c. $\overline{A} \cup \emptyset = \emptyset$?

(Hint: always, sometimes, never)

d. Why does $A \cap U = A$?

6. Verify the commutative law for intersection for the sets

$A = \{\triangle, 0, \square, \diamond\}$

$B = \{\triangle, 0, 11\}$

7. True or false?

a. $A \cap B = A$ if $A \subseteq B$

b. $A \cap \overline{A} = U$

c. $A \cup (B \cap C) = (A \cup B) \cap (A \cup C)$

d. If $C \subset A$ and $C \subset B$, then $A \subseteq (C \cap B)$

8. Given

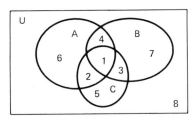

FIGURE 3-11

where $A = \{1, 2, 4, 6\}$, $B = \{1, 3, 4, 7\}$, $C = \{1, 2, 3, 5\}$.

Find a. $\{1, 2, 4, 6\} \cup \{1, 2, 3, 5\}$

  b. $\{1, 3, 4, 7\} \cap \{1, 2, 4, 6\}$

  c. $A \cap B \cap C$

  d. $\underline{A} \cup (B \cap C)$

  e. $\overline{A} \cap (B \cup C)$

9. Mr. Jones owns an automobile parts warehouse. He has the option of buying, at discounts, the following items from dealers A, B, and C:

Dealer A   headlights

Dealer B   fan belts, headlights, spark plugs

Dealer C   headlights, spark plugs, motor oil

Use the second distribution law to illustrate buying only the headlights and spark plugs.

4

LOGIC

In mathematical logic we use exact terms in very special and restricted ways. Since it is restricted and exact, mathematical logic has many direct applications to computer decision making. We will be discussing computer decision making in Chapter 6. First, however, we need to understand exactly what we mean when we speak of logic in mathematical terms.

When an event occurs several times, we begin to think that there is a reason for it. We look for a pattern. When the ancient Babylonians observed the motion of the stars across the sky, they were looking for a pattern. The builders of Stonehenge on Salisbury Plain in England dramatically demonstrated the patterns they found in the seasons of the year.

Looking for patterns is called *inductive reasoning*. Finding the patterns and trying to prove them is called *deductive reasoning*.

Inductive reasoning may result in conclusions that can only be considered probable. Deductive reasoning must result in a valid conclusion provided the input information is true. Deductive reasoning is the approach the computer programmer must take. Input information must be correct. The output required must be reliable. The procedures linking inputs and outputs comprise the logical sequence of instructions to the computer, written by the programmer.

Deductive reasoning is the basis of mathematical proof. It is frequently presented in conditional forms such as, if the given information is true, then the conclusion must be true.

Statements that make up the beginning of an argument are true, or assumed to be true, and are referred to as the *hypothesis*.

Statements that make up the goal of the argument are called *conclusions*.

Logic is the study of the rules used to operate on an hypothesis or several hypotheses and reach a conclusion. As in Chapter 3, we will use Venn diagrams to make our discussion clear. First, we must outline the exact terms and rules mentioned in our opening paragraph. Deductive reasoning requires that we

1. accept the hypothesis as being true, and
2. agree upon the rules, or logic, used to reach conclusions.

We must understand the restrictions on the type of statements we can make. We will use only declarative sentences that can be *true* or *false*. This limiting to two possibilities is comparable to the binary use of 0 and 1. Ambiguous or nonsensical statements are not acceptable. Also, in our discussions we will *not* use:

1. interrogative statements that ask a question, e.g., What time is it?
2. exclamatory statements that express emotion, e.g., How beautiful the garden looks!
3. imperative sentences that give a command, e.g., Turn off that motor!

To study the rules of logic we will use *declarative sentences* that are either true or false, but not both.

Although they do not constitute a proof, Venn diagrams can be used with sets to illustrate simple arguments. Consider the following:

If it is a dog, then it is an animal.

Let *U* represent the universe of items under discussion. If circle *A* represents all animals and circle *D* represents all dogs, a proper subset of animals, then the Venn diagram illustrates the statement. The Venn diagram also shows that dogs are represented by all points inside *D* and all animals that are not dogs are represented by all points outside *D* but inside *A*. (See Fig. 4-1.)

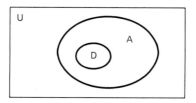

FIGURE 4-1

The Venn diagram can be used to demonstrate set relationships characterized by the terms *all, none,* and *some.*

4-2 THE TERM "ALL"

Consider an employee file containing records for all of the employees for a computer service bureau. Figure 4-2 can be used to illustrate the statement "*All* Computer programmers are data processing employees." Other ways of describing *all* can be seen in Fig. 4-3. In a public school district there are high schools, junior high schools, and elementary schools. The elementary school contains the following grades: kindergarten, first grade, second grade, third grade, fourth grade, fifth grade, and sixth grade. *All* of these seven grades are members of the subset *elementary schools.* Further, the largest enclosed shape in Fig. 4-3 includes *all* of the schools in the district. In a general way, we can state that if all A is in B, then A is a subset of B.

4-3 THE TERM "NONE"

Consider a deck of playing cards; some have red markings and some have black. In Fig. 4-4, let B represent the set of black cards and R represent the set of red cards; then B and R are circles which contain no points in common and make up disjoint sets.

The statements
1. has black markings
2. has red markings

cannot be true for the same card. We could restate our two premises thus:

1. If it is black, then it is not red.
2. If it is red; then it is not black.

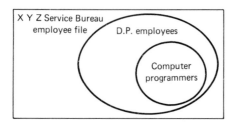

FIGURE 4-2

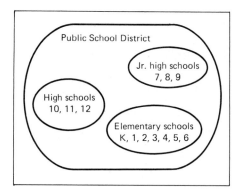

FIGURE 4-3

Reasoning or inference occurs whenever we assert something to be true on the basis of something else being true. Figure 4-5 shows a Venn diagram illustrating a combination of events for *all* and *none*. If we let

$$K = \text{keypunch operators}$$
$$E = \text{employees}$$
$$C = \text{computers}$$

we can illustrate simple deductive reasoning as follows:
Hypotheses:
Premise 1: No C's are E's.
Premise 2: All K's are E's.
The following *conclusion* is deducible:
Conclusion: No C's are K's.
None, in this example means that no elements of C are elements of K or E. (The intersection of set C with set K or set E yields an empty or null set.)

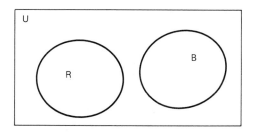

FIGURE 4-4

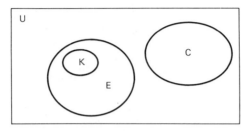

FIGURE 4-5

4-4 THE TERM "SOME"

The term "some," like "all" and "none," has a very precise meaning. In logic, "some" means "at least one, but not all." The fact that *some* elements are present in both of two sets means that intersection exists for the two sets, yielding a non-empty set.

The statement "some of the ABC Department Store's customers have credit balances" can be expressed in the Venn diagram shown in Fig. 4-6

where

U = all things in the ABC Department Store's accounting system
A/R = accounts receivable (or customers in this case)
A/P = accounts payable, or accounts to which the department store owes money or credit.

The shaded intersection of the two enclosed areas contains those customers with credit balances, or those accounts receivable to whom the department store owes money or credit.

A statement implied in Fig. 4-6 is:

Some accounts payable are not customers.

The points in A/R that are not shaded include customers that are not

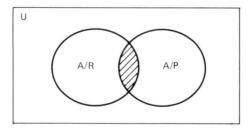

FIGURE 4-6

owed money (and considered as an account due), and the points in *A/P* that are not shaded include accounts payable that are not customers.

_____ **4-5 ARGUMENTS**

Reasoning or inference that is intended to convince someone of the truth of an assertion is usually called an *argument.*

Definition 4-1

An *argument* is a group of at least three statements, including two or more *premises* and a conclusion that is obtained from the premises.

Arguments are usually given in a step-by-step list of statements beginning with the hypotheses or premises, and ending with the conclusion.

EXAMPLE 1

Hypotheses:—All firemen are city employees.
　　　　　　—All city employees are high school graduates.
　Conclusion:—All firemen are high school graduates.
Example 1 illustrates an argument by Definition 4-1.

EXAMPLE 2

Statement 1—All firemen are city employees.
Statement 2—All city employees are high school graduates.

Example 2 does not represent an argument because it does not contain a conclusion.

Another important rule is that we accept the hypotheses as true. This has already been stated, but in the following examples we will show this restriction can force conclusions which would normally be rejected as false. Remember, we are trying to establish not the truth of the original statement, but the rules of logic. In general:

1. An argument is *valid* if and only if its conclusion is derivable from its premises.
2. An argument is *invalid* if and only if its conclusion is not derivable from its premises.

This linking of the validity of an argument to the reasoning involved, instead of to the "truth" of the hypotheses, is a key point.

The following three examples are given so that the truth or falsity of a hypothesis may not be confused with the validity or invalidity of an argument.

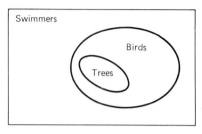

FIGURE 4-7

I. Hypotheses:

> All trees are birds.
> All birds can swim.

Conclusion:

> All trees can swim.

Although the hypotheses are false, the argument is valid (Fig. 4-7).

Another example illustrates an invalid argument with a true conclusion.

II. Hypotheses:

> All females sing.
> All women sing.

Conclusion:

> All women are females.

Let

$$S = singers$$
$$F = females$$
$$W = women$$

The Venn diagram shown in Fig. 4-8 satisfies the hypotheses, but does not force the conclusion; therefore, the argument is invalid. Note that the validity of the argument has nothing to do with the obvious truth of the conclusion.

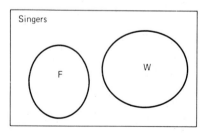

FIGURE 4-8

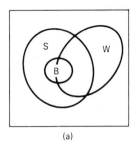
(a)

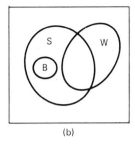
(b)

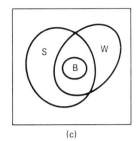
(c)

FIGURE 4-9

Consider this use of the term *some:*

III. Hypotheses:

Some students are witty.
All boys are students.

Conclusion:

All boys are witty.

If boys are a proper subset of students, more information is needed before we can conclude that all boys are witty. Let

$$S = \text{students}$$
$$W = \text{witty}$$
$$B = \text{boys}$$

Under these hypotheses, three Venn diagrams are possible. These diagrams are shown in Fig. 4-9.

The argument is invalid because two of the three Venn diagrams (*a* and *b*) do not force the conclusion that all boys are witty.

Exercises 4-1

Using the terms *all*, *none*, and *some*, describe the following Venn diagrams.

EXAMPLE

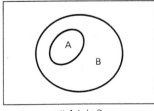

all A is in B

FIGURE 4-10

1.

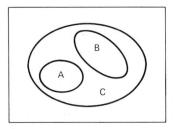

FIGURE 4-11

2.

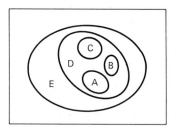

FIGURE 4-12

3.

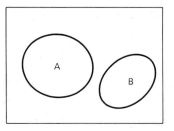

FIGURE 4-13

4.

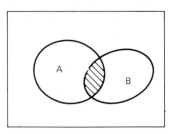

FIGURE 4-14

5.

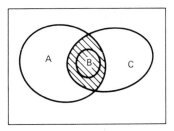

FIGURE 4-15

6.

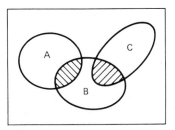

FIGURE 4-16

7.

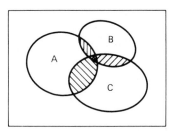

FIGURE 4-17

8. Given the following table of rules for valid arguments, state whether the arguments that follow are valid or invalid. Justify your answer by drawing the appropriate Venn diagram.

Hypothesis	Conclusion	Result
True	True	Valid
True	False	Invalid
False	True	Valid
False	False	Valid

In exercises (a), (b), (d), (e) and (g) assume the hypotheses to be true.

a. Hypotheses: All *A* is *B*
All *B* is *C*
   Conclusion: All *A* is *C*
b. Hypotheses: All *X* is *Y*
All *Z* is *Y*
   Conclusion: All *X* is *Z*
c. Hypotheses: No teachers are students.
All students are witty.
   Conclusion: No teachers are witty.
d. Hypotheses: Some *A*'s are *B*'s
All *C*'s are *A*'s
   Conclusion: All *C*'s are *B*'s
e. Hypotheses: All programmers are people.
All people are fallible.
   Conclusion: All programmers are fallible.
f. Hypotheses: All calculators are computers.
All computers have terminals.
   Conclusion: All calculators have terminals.
g. Hypotheses: All *E*'s are *F*'s
Some *F*'s are *G*'s
   Conclusion: Some *E*'s are *G*'s
   (Is more than one Venn diagram possible here?)

9. Give an example of your own in which more than one Venn diagram is possible.
10. What is the difference between inductive and deductive reasoning?
11. Identify the declarative sentences that are not nonsensical.

a. Iowa is one of fifty states.
b. Cats grow on trees.
c. $x - .5 = 3$.
d. Snow melts.
e. There is a counting number x such that $1/0 = x$.

4-6 COMPOUND STATEMENTS AND TRUTH TABLES

Definition 4-2

A compound statement is a statement formed by combining two or more simple statements. (Again, declarative statements are used; see page 83.) Consider, for example:

It is Friday.
The checks are out.

It is possible to combine these two simple declarative statements by using the connective *and* or the connective *or.*

1. It is Friday *and* the checks are out.
2. It is Friday *or* the checks are out.

Using a lower case alphabetic character as a placeholder for each simple statement, we can substitute as follows.

$$\text{Let } \quad p = \text{It is Friday.}$$
$$q = \text{The checks are out.}$$

Using the symbols $\wedge$ for the connective *"and"* and $\vee$ for the connective *"or,"* we can now write the following two compound statements:

Compound Statement 1) $p \wedge q$: It is Friday and the checks are out.
Compound Statement 2) $p \vee q$: It is Friday or the checks are out.

$\vee$ means the *inclusive* use of *or*; i.e., either or both. When it is necessary for the *exclusive* use of *or,* we will mean either, *but not both,* and use the symbol $\underline{\vee}$.

4-6.1 Conjunction Statement

A sentence containing "and" is a conjunction of two simple statements. It is expressed in the form of *"p and q"* and is written symbolically as $p \wedge q$.

Remembering that statements in logic can be true or false, but not both, we will consider all the possibilities for $p \wedge q$.

p is true and q is true.
p is true and q is false.
p is false and q is true.
p is false and q is false.

The connective *and* suggests that a compound statement is true when the simple statements are *both* true.

For example, in the statement

$$2 < n \text{ and } n > 2$$
$$\text{let } \quad p = 2 < n$$
$$q = n > 2.$$

Consider $n \in R$, i.e., n represents any real number and let T represent true and F represent false.

When p is T and q is T, then $p \wedge q$ is T.
When p is T and q is F, then $p \wedge q$ is F.
When p is F and q is T, then $p \wedge q$ is F.
When p is F and q is F, then $p \wedge q$ is F.

Table 4-1

p	q	p ∧ q
T	T	T
T	F	F
F	T	F
F	F	F

These four possibilities can be grouped into what is called a truth table for *conjunction* (Table 4-1).

Note that the conjunction of two statements is true only when both statements are true.

4-6.2 Disjunction Statement

A sentence containing *or* is a disjunction of two simple statements. It is expressed in the form of "*p* or *q*" and is written symbolically as $p \lor q$. The disjunction of two statements is true if either one or both of the statements is true. We see this in Table 4-2, the truth table for disjunction.

In the statement "*x* is a prime number or *x* < 6," let *p* = the number *x* is a prime and *q* = *x* < 6 (where *x* is a natural number). Then all the possibilities for $p \lor q$ can be expressed in the truth table for disjunction.

4-6.3 Negation Statement

There is another possibility we need to discuss, namely, the term "not." Earlier, we had the set $\overline{A}$, which included everything in the universal set, but not in *A*. We have a similar, but not identical, case in logic. The negation symbol is ~. If we let *p* represent the statement "the highway is wet," then ~*p* means "The highway is not wet." Since we have made the restriction that all statements must be true or false, if *p* is true then ~*p* must be false. Table 4-3 shows the two possibilities for negation.

Table 4-2

p	q	p ∨ q
T	T	T
T	F	T
F	T	T
F	F	F

Table 4-3

p	$\sim p$
T	F
F	T

It is possible to construct a truth table combining the three symbols used so far Consider

$$(p \wedge q) \vee \sim p$$

p	q	$(p \wedge q)$	$\sim p$	$(p \wedge q) \vee \sim p$
T	T	T	F	T
T	F	F	F	F
F	T	F	T	T
F	F	F	T	T

To construct this table, we first enter the values for p and q and complete the column for conjunction $(p \wedge q)$. Next enter values for $\sim p$ by negating the values under the column headed p. The columns headed $(p \wedge q)$ and $\sim p$ represent, respectively, the values for p and q in Table 4-2. Using the table for disjunction, we complete the column headed $(p \wedge q) \vee \sim p$.

Exercises 4-2

1. In the following sentences pick out the simple statements, assign lower case letters to them, and rewrite the entire sentence in symbolic notation.
    a. It is raining and the lawn is wet.
    b. *A* is here or *C* is lost.
    c. Dogs are loud and cats are quiet.
    d. Dogs are loud or cats are not quiet.
    e. $z = 10$, and $y = 9$, and $x = 8$.
2. Given p = Cows wear bells.
    q = Crows steal corn.
    Write the following in words:
    a. $p \vee q$ e. $\sim p \vee (p \wedge q)$
    b. $p \wedge q$ f. $(\sim p \wedge q) \vee (p \wedge \sim q)$
    c. $p \vee \sim q$ g. $\sim \sim q$
    d. $\sim p \wedge \sim q$
3. Write the truth tables for a, c, and e in Exercise 2.
4. The truth table for disjunction describes the *inclusive* meaning of *or*. If

$p \vee q$ is defined as p or q but not both, construct the truth table for this *exclusive* meaning of *or*.

5. Two statements are equivalent if their truth values are the same. Are the following equivalent statements? (Construct the truth tables.)

   a. $\sim p \vee \sim q$ and $\sim (p \vee q)$
   b. $\sim p \wedge \sim q$ and $\sim (p \wedge \sim q)$

6. The prerequisite for a position as a Cobol programmer trainee is: A two-year degree in data processing. One year's experience in programming in Cobol is acceptable but not necessary.

   Write the truth table for this situation.

7. Prerequisites for a position as a Fortran programmer are:

   1. two-year college degree in data processing AND
   2. one year's experience in programming OR
   3. three years' experience as a Fortran programmer.

   a. Complete the truth table below for this situation. (Note: Three "placeholders," p, q, and r are required for the three simple statements above. Hint: There are 2^n entries needed in any truth table, where n is the number of placeholders.

p	q	r	
T	T	T	
T	T	F	
T	F	T	
T	F	F	etc.
F	T	T	
F	T	F	
F	F	T	
F	F	F	

   b. Construct the truth table for (p OR q) AND r

4-7 OTHER TYPES OF COMPOUND STATEMENTS

We noted earlier that many statements about sets were characterized by the words *all, none,* and *some*. Many statements about logic are characterized by the words *if* and *then*.

4-7.1 Implication Statement

A sentence of implication is a conditional statement. It is found in the form "If . . . , then . . ." and can be expressed symbolically as $p \rightarrow q$, which is read "if *p*, then *q*" or "*p* implies *q*," where *p* and *q* are two simple statements. The symbol → is the implication symbol and is the connective used to form this type of compound statement. Calling the "if clause" the *antecedent* and the "then clause" the *consequent,* we can name the two parts of the conditional statement "If *A* is a motorcycle, then *A* has two wheels:

Antecedent: $p = A$ is a motorcycle.
Consequent: $q = A$ has two wheels.

We will examine the four possible truth values of: if *p* then *q*, or $p \rightarrow q$.

1. If the antecedent is true and the consequent is true, then the statement $p \rightarrow q$ is true.
2. If the antecedent is true and the consequent is false then the statement $p \rightarrow q$ is false.
3. If the antecedent is false (or *A* is not a motorcycle) and the consequent is true then $p \rightarrow q$ is true. *A* may be any other two-wheeled vehicle.
4. If the antecedent is false and the consequent is false then $p \rightarrow q$ is true.

A summary of this analysis is shown in Table 4-4, the truth table for conditional statements. We see from this that $p \rightarrow q$ is false only when *p* is true and *q* is false. In every other case it is true.

In our daily conversations, conditional sentences are typically used in a meaningful way. However, it should be noted that, under our rules, it is also possible to have antecedents and consequents such as

If today is Tuesday, then $A > B$.
If $5 > 2$, then the moon lander is in orbit.

Further, these conditional statements will be true or false in accordance with how they are defined in Table 4-4.

4-7.2 Biconditional Statement

When we state both "If *p*, then *q*" and "if *q*, then *p*," we use a sentence with "double implication," or a biconditional statement. A bicondi-

Table 4-4

p	*q*	$p \rightarrow q$
T	T	T
T	F	F
F	T	T
F	F	T

tional statement is symbolized as $p \leftrightarrow q$, where $p \leftrightarrow q$ means the same as $(p \rightarrow q) \land (q \rightarrow p)$.

In p: The triangle is an equilateral triangle.

   q: The triangle has three equal angles.

The biconditional statement, $p \leftrightarrow q$, is read "The triangle is an equilateral triangle if and only if the triangle has equal angles."

Using the truth tables for conjunction statements and conditional statements, we can derive the biconditional truth table.

Given

p	q	$p \rightarrow q$		q	p	$q \rightarrow p$
T	T	T		T	T	T
T	F	F	and	F	T	T
F	T	T		T	F	F
F	F	T		F	F	T

then

$p \rightarrow q$	$q \rightarrow p$	$(p \rightarrow q) \land (q \rightarrow p)$
T	T	T
F	T	F
T	F	F
T	T	T

From these two truth tables, we can find $(p \rightarrow q) \land (q \rightarrow p)$. Since $(p \rightarrow q) \land (q \rightarrow p)$ means the same as $p \leftrightarrow q$, we then have the truth table for the biconditional statement, Table 4-5.

If either, but not both, p or q is false, then $p \leftrightarrow q$ is false. If p and q are both true or both false, then $p \leftrightarrow q$ is true.

Table 4-5

p	q	$p \leftrightarrow q$
T	T	T
T	F	F
F	T	F
F	F	T

Two statements are equivalent if and only if they have identical values in the last column of their truth tables. For example, $p \rightarrow q$ and $\sim q \rightarrow \sim p$ are equivalent statements.

p	q	$p \rightarrow q$		$\sim q$	$\sim p$	$\sim q \rightarrow \sim p$
T	T	T		F	F	T
T	F	F		T	F	F
F	T	T		F	T	T
F	F	T		T	T	T

From these two truth tables, we see that the last columns are identical; hence the two statements $p \rightarrow q$ and $\sim q \rightarrow \sim p$ are equivalent statements.

4-7.3 Other Forms of the Conditional Statement

Given $p \rightarrow q$, three other related statements are possible.

1. The converse, $q \rightarrow p$: If q, then p.
2. The inverse, $(\sim p) \rightarrow (\sim q)$: If not p, then not q.
3. The contrapositive, $(\sim q) \rightarrow (\sim p)$: If not q, then not p.

The conditional sentence "If the sun is shining, then I wash cars" serves to illustrate these three related forms:

Converse: If I wash cars, then the sun is shining.
Inverse: If the sun is not shining, then I do not wash cars.
Contrapositive: If I do not wash cars, then the sun is not shining.

EXAMPLE   Write the converse, inverse and contrapositive for the statement "If $a^2 - b^2 = 1$, then $(a + b)(a - b) = 1$."

Converse: If $(a + b)(a - b) = 1$, then $a^2 - b^2 = 1$.
Inverse: If $a^2 - b^2 \neq 1$, then $(a + b)(a - b) \neq 1$.
Contrapositive: If $(a + b)(a - b) \neq 1$, then $a^2 - b^2 \neq 1$.

A statement and its converse, or a statement and its inverse, do not always have the same truth values. This can be seen in Table 4-6. Consider the statement "If we run, then we will tire." Accepting this statement as true, study its converse: "If we tire, then we ran." (Let p = We run; q = We will tire.)

Being tired is not necessarily a result of running. Likewise, the inverse, "If we do not run, we will not tire, is also not necessarily true.

The contrapositive, "If we do not tire, then we did not run," is equivalent to the original statement: if you did not tire, you could not have been running.

The following truth table summarizes all the possibilities for the three variations of the conditional statements we have been discussing. Do you see how the columns for the converse, inverse, and contrapositive were obtained? Note that the contrapositive of a statement may be formed by the converse of the inverse, or the inverse of the converse, of the original statement.

Table 4-6

		Statement	Converse	Inverse	Contrapositive
p	q	$p \rightarrow q$	$q \rightarrow p$	$(\sim p) \rightarrow (\sim q)$	$(\sim q) \rightarrow (\sim p)$
T	T	T	T	T	T
T	F	F	T	T	F
F	T	T	F	F	T
F	F	T	T	T	T

_____ **4-8 TAUTOLOGIES**

Definition 4-3

A tautology is a compound declarative sentence that is true regardless of the truth values assigned to its individual component statements.

A tautology is identified by all *T*'s in its final column in a truth table. Tautologies are basic tools in determining the validity of arguments. For example, consider the truth table for $p \rightarrow (p \vee q)$.

Table 4-7

p	q	$p \vee q$	$p \rightarrow (p \vee q)$
T	T	T	T
T	F	T	T
F	T	T	T
F	F	F	T

Under p and q enter the values for p and q as before. Under $p \vee q$ list the values found in the truth table for disjunction. Combine the columns under p and $p \vee q$ using the conditional truth table to form the last column $p \rightarrow (p \vee q)$. Because there are no *F*'s in the last column, $p \rightarrow (p \vee q)$ is a tautology. If we change the symbol $\vee$ to $\wedge$ then $p \rightarrow (p \wedge q)$ is shown in Table 4-8.

Table 4-8

p	q	$p \wedge q$	$p \rightarrow (p \wedge q)$
T	T	T	T
T	F	F	F
F	T	F	T
F	F	F	T

The truth table for $p \rightarrow (p \wedge q)$ contains one F in its final column; therefore, $p \rightarrow (p \wedge q)$ is not a tautology.

4-9 CONTRADICTIONS

When the last column contains all F's we have a *contradiction*.

Table 4-9

p	q	$(p \vee q)$	$\sim(p \vee q)$	$(p \vee q) \leftrightarrow \sim(p \vee q)$
T	T	T	F	F
T	F	T	F	F
F	T	T	F	F
F	F	F	T	F

Consider the truth table for $(p \vee q) \leftrightarrow \sim(p \vee q)$. Because the last column contains all F's, $(p \vee q) \leftrightarrow \sim(p \vee q)$ is called a contradiction. (In this last example $(p \vee q)$ was treated as the p and $\sim(p \vee q)$ was treated as the q in the biconditional truth table.)

Exercises 4-3

1. Rewrite the following in the form if . . . , then . . . ,
   a. Equilateral triangles have equal sides.
   b. Friday is payday.
   c. A stitch in time saves nine.
2. Which of the following are conditional statements and which are biconditional?
   a. If it is a bird, then it flies.
   b. $y = mx + b$ if and only if $b = y - mx$.

  c. If it is a quadrilateral, then it is a polygon.
  d. If it is snowing, the temperature is below 33°.
  e. $x^2 - x - 6 = 0$ if and only if $x = 3$.

3. Write the examples in Exercise 2 in symbolic notation.
4. Given: $p = A$ is tall
  $q = B$ is short
  $r = C$ is medium
  Write the English equivalents of

  a. $p \rightarrow q$ f. $(q \wedge r) \rightarrow p$
  b. $p \leftrightarrow r$ g. $(q \vee r) \leftrightarrow (r \vee q)$
  c. $p \rightarrow q \wedge q \rightarrow p$ h. $\sim p \rightarrow r \vee (q \vee r)$
  d. $q \rightarrow r$ i. $(p \rightarrow q) \leftrightarrow (\sim r \vee \sim q)$
  e. $\sim p \rightarrow \sim q$ j. $p \rightarrow (\sim p \rightarrow q)$

5. Construct a truth table for b and j in Exercise 4.
6. a. Is $p \leftrightarrow (p \vee q)$ a tautology?
  b. Construct a contradiction in symbolic notation and prove it by its associated truth table.
7. Write the converse, inverse, and contrapositive for

  a. $p \rightarrow q$
  b. $p \rightarrow \sim q$

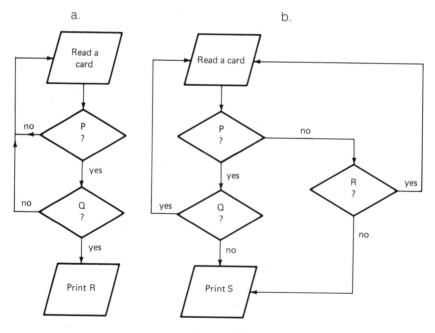

FIGURE 4-18

8. Write an equivalent statement for $(p \rightarrow q) \wedge (q \rightarrow p)$. Construct the truth tables to prove your answer.

9. Verify the following tautologies by constructing their truth tables.

   a. $p \wedge q \rightarrow p$
   b. $[p \wedge (p \rightarrow q)] \rightarrow q$
   c. $\sim(p \wedge q) \leftrightarrow \sim p \vee \sim q$
   d. $\sim(p \vee q) \leftrightarrow \sim p \wedge \sim q$
   e. $\sim(p \rightarrow q) \leftrightarrow p \wedge \sim q$

10. Describe the activities of the flowcharts in Fig. 4-18 and express these activities by writing the appropriate statements in symbolic notation for printing.

5

BOOLEAN ALGEBRA
AND SWITCHING
NETWORKS

──────────── 5-1 ELEMENTS IN BOOLEAN ALGEBRA

We are interested in Boolean Algebra because it relates to the networks in a computer system. Boolean Algebra and computer logic are often associated. Historically, Boolean Algebra was related to the formal study of logic by George Boole in 1854. Claude Shannon, in 1938, applied Boolean principles to switching networks.

An electrical circuit contains many elements: switches, resistances, direction, and so on. However, only the switches are of importance in Boolean Algebra.

Consider a simple light switch in a kitchen. If a light switch is open, electricity cannot flow through the circuit and the light is off. When the light switch is closed and electricity does flow through the circuit, the light is on. (Refer to Figs. 5-1 and 5-2.)

Switches can only exist in one of two states: on or off. When a switch is closed so that current can pass through, it is in the on-condition, and we give it the value 1. When a switch is opened so that no current can pass through, it is in the off-condition, and we give it the value 0. A single wire going from a power source, such as a battery, to an output station, such as a light bulb, and back to the power source can also be described as having a value of 1. (See Fig. 5.3) To understand the logic of this, we could consider the circuit to have a switch in the permanent on-condition.

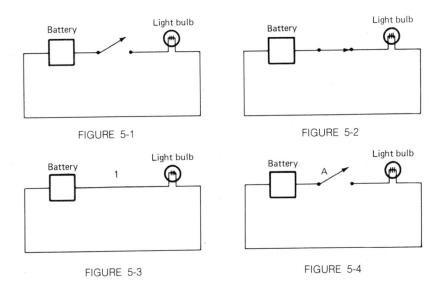

FIGURE 5-1

FIGURE 5-2

FIGURE 5-3

FIGURE 5-4

In Fig. 5-4, a switch is inserted in the circuit. Now the on/off condition of the light bulb depends upon the on/off condition of switch *A*.

A network or circuit is described in this chapter as a combination of connected switches, wires, and input and output stations. Our schematic reference to a switch will be

*i*_____*A*_____*o*, where

    A is a variable representing a switch
    i means input
    o means output (to be distinguished from the ellipse 0, meaning zero

By letting 1 represent the on-condition and 0 represent the off-condition of a switching network, it is possible to apply Boolean Algebra to switching networks.

_____ 5-2 AXIOMS OF BOOLEAN ALGEBRA

We will refer to these two conditions of an electrical circuit as elements in the two-element Boolean Algebra applied to computers. Because the complete set of elements to be operated on is $B = \{0, 1\}$, we are concerned with the following properties:

Definition 5-1

The class of elements *B* is a Boolean Algebra if and only if the following axioms hold:

1. Addition ($+$) and multiplication ($\cdot$) are the only permitted binary operations in *B*.

2. The operations (+) and (·) are commutative for all elements in *B*.
3. Each operation is distributive over the other.
4. There exist only two identity elements in *B*: 0 and 1.
5. There exists an element *A* in *B*, read "a complement of *A*," such that $A + \overline{A} = 1$ and $A \cdot \overline{A} = 0$.

It can be shown that the following are characteristics of a two-element Boolean Algebra:

1. No negative number is possible.
2. All variables exist to the zero or first power.
3. There are no roots.

These characteristics will prove to be important when we discuss the basic properties of circuits.

To better understand how Boolean Algebra provides a model for computer circuits, remember the following three associations: addition is associated with a *parallel* circuit; multiplication is associated with a *series* circuit; *complementation* refers to the change in the condition of a switch from on to off or from off to on. (Recall that *B* = {0, 1}.)

When two switches are connected in parallel, the circuit is in the on-condition when either or both switches are closed. The circuit is in the off-condition only when both switches are open. Figure 5-5 shows a parallel circuit.

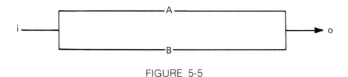

FIGURE 5-5

When two switches are connected in series, the circuit is in the on-condition only when both switches are closed. The circuit is in the off-condition when either or both switches are open. Figure 5-6 shows a series circuit.

FIGURE 5-6

5-3 ADDITION AND THE PARALLEL CIRCUIT

In Fig. 5-5, if a pulse of electricity is available at *i*, then either one or both of the switches *A* and *B* must be closed to produce output at *o*. The

Boolean operation $A + B$ is like turning on one light from either one of two switches or both switches. Since either or both switches may be open or closed at any given time, there are four possible events in Fig. 5-5.

In Table 5-1, 0 represents an open switch and 1 represents a closed switch. This table shows the relationship between Boolean addition and the parallel circuit: the result or output of a parallel circuit can only exist in one of two conditions, on or off. (Note that this addition table is similar to the truth table for disjunction, Table 4-2, where $T = 1$ and $F = 0$.)

5-4 MULTIPLICATION AND THE SERIES CIRCUIT

In Fig. 5-6, if a pulse of electricity is available at i, then both switch A and switch B must be closed to produce output at o. Again, as with the two switches in parallel, there are four possible events. These possibilities are shown in Table 5-2. (Note that this table is similar to the truth table for conjunction, Table 4-1, where $T = 1$ and $F = 0$.)

Table 5-1

A	B.	A + B	Result
1	1	1 + 1	1
1	0	1 + 0	1
0	1	0 + 1	1
0	0	0 + 0	0

Table 5-2

A	B	A · B	Result
1	1	1 · 1	1
1	0	1 · 0	0
0	1	0 · 1	0
0	0	0 · 0	0

5-5 COMPLEMENTS

In some networks two switches may always exist in opposite conditions. When the first switch is on, the second is off, and when the second is on, the first is off. It is possible to express this situation in Boolean notation by placing a bar over a variable: $\overline{A}$ means not A.

If the variable represents the condition of a certain switch A, then $\overline{A}$ means the opposite of that condition.

If A is on, then $\overline{A}$ represents the off-condition;
If A is off, then $\overline{A}$ represents the on-condition.

$\overline{A}$ is the complement of A, and in our networks, A and $\overline{A}$ may represent two switches. A complete set of events for switch A consists of two electrical conditions (Table 5-3).

Table 5-3	
A	$\overline{A}$
1	0
0	1

Table 5-4			
A	$\overline{A}$	$A + \overline{A}$	Result
1	0	1 + 0	1
0	1	0 + 1	1

This table is similar to negation in logic statements, where $T = 1$ and $F = 0$. The network for $A + \overline{A}$ can be drawn in a parallel arrangement as shown in Fig. 5-7.

All possible states of electricity for this network are shown in Table 5-4. The result for $A + \overline{A}$ will always be 1. That is, electricity will always flow through the circuit of Fig. 5-7.

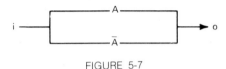

FIGURE 5-7

Since $\overline{A}$ represents the opposite condition of an existing variable and is not a new variable, Table 5-4 has only two possible results.

The network for $A \cdot \overline{A}$ can be drawn in a series arrangement. This is shown in Fig. 5-8. Table 5-5 shows all the possible states of electricity for this network.

FIGURE 5-8

The result for $A \cdot \overline{A}$ is always 0. This means that under no conditions will current flow in this series circuit.

We have studied the relationship between series circuits and Boolean multiplication, and parallel circuits and Boolean addition. Now we will see how multiplication and addition can be combined to help us investigate circuits that have switches in both series and parallel.

Figure 5-9 shows a network which we can represent as $B \cdot (A + \overline{B})$. The four possible conditions of this network are listed in Table 5-6.

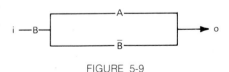

FIGURE 5-9

Table 5-5

A	$\overline{A}$	$A \cdot \overline{A}$	Result
1	0	$1 \cdot 0$	0
0	1	$0 \cdot 1$	0

Table 5-6

A	B	$\overline{B}$	$B \cdot (A + \overline{B})$	Result
1	1	0	$1(1 + 0)$	1
1	0	1	$0(1 + 1)$	0
0	1	0	$1(0 + 0)$	0
0	0	1	$0(0 + 1)$	0

In this case, even though we have three switches, there are only four possible conditions. This is because the symbol $\overline{B}$ does not represent a new variable and we are only dealing with two variables (A and B). Therefore, $B \cdot (A + \overline{B})$ has 2^2 or 4 possible conditions (see page 68, Ex. 9).

Exercises 5-1

1. Name the five basic properties of Boolean Algebra.
2. Refer to Fig. 5-10 to answer the following

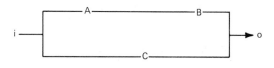

FIGURE 5-10

Will current flow through to o when
   a. $A = 1, B = 1, C = 0$?
   b. $A = 0, B = 1, C = 0$?
   c. $A = 1, B = 1, C = 1$?
   d. $A = 0, B = 0, C = 1$?
3. The complete set of conditions (elements) to be operated on in Boolean Algebra are 0 and 1; name the two operations permitted.
4. Draw the circuit for each of the following and write a table of the appropriate on/off conditions and results.
   a. $A + B + C$
   b. $A \cdot B \cdot C$
5. What does "$\overline{A}$ is the complement of A" mean?
6. Given: $A + \overline{A}B$.
   a. Draw the appropriate circuit.
   b. Write the corresponding table.
7. If a network contains 3 variables, how many possible results exist?

8. Given:

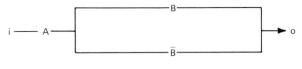

FIGURE 5-11

   a. Write the Boolean expression for Fig. 5-11.
   b. Is output at o dependent upon B, $\overline{B}$ or A?

9. Complete this table:

A	$\overline{A}$	B	$A \cdot B + \overline{A}$	Result
1	0	1	$1 \cdot 1 + 0$	
1	0	0		
0	1	1		
0	1	0		

10. Complete this table:

A	B	$\overline{B}$	$B \cdot ($	Result
1	1	0		1
1	0	1		0
0	1	0		0
0	0	1		0

_____ 5-6 COMBINATIONS OF SWITCHES

   Combinations of switches are often found in series-parallel networks. Consider the network in Fig. 5-12. Switch A is in series with parallel switches B and C. We have isolated this part of the network in Fig. 5-13.

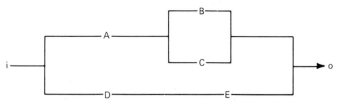

FIGURE 5-12

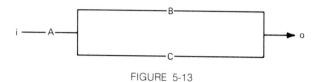

FIGURE 5-13

Also switches D and E are in series (Fig. 5-14).

The Boolean statement of the network shown in Fig. 5-13 is $A \cdot (B + C)$, and the Boolean statement of the network shown in Fig. 5-14 is $D \cdot E$. If we put the network of Fig. 5-13 in parallel with that of Fig. 5-14, we obtain the original network in Fig. 5-12. Therefore, the Boolean statement for the total network is:

$$[A \cdot (B + C)] + (D \cdot E)$$

Table 5-7 lists all the possible events for the network shown in Fig. 5-13.

Table 5-7

A	B	C	$A \cdot (B + C)$	Result
1	1	1	$1 \cdot (1 + 1)$	1
1	1	0	$1 \cdot (1 + 0)$	1
1	0	1	$1 \cdot (0 + 1)$	1
1	0	0	$1 \cdot (0 + 0)$	0
0	1	1	$0 \cdot (1 + 1)$	0
0	1	0	$0 \cdot (1 + 0)$	0
0	0	1	$0 \cdot (0 + 1)$	0
0	0	0	$0 \cdot (0 + 0)$	0

Note that with three variables there are $2^3 = 8$ possible conditions. Four variables yields $2^4 = 16$ possible conditions. There are always 2^n possible conditions, where n is the number of different variables in a circuit. As an exercise, write the table for Fig. 5-12.

Given $D \cdot (A \cdot B + C)$, we can draw the network shown in Fig. 5-15. All sixteen possible conditions for this network are given in Table 5-8.

FIGURE 5-14

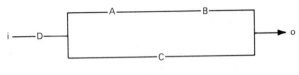

FIGURE 5-15

Table 5-8

A	B	C	D	$D \cdot (A \cdot B + C)$	Result
1	1	1	1	1(1 · 1 + 1)	1
1	1	1	0	0(1 · 1 + 1)	0
1	1	0	1	1(1 · 1 + 0)	1
1	1	0	0	0(1 · 1 + 0)	0
1	0	1	1	1(1 · 0 + 1)	1
1	0	1	0	0(1 · 0 + 1)	0
1	0	0	1	1(1 · 0 + 0)	0
1	0	0	0	0(1 · 0 + 0)	0
0	1	1	1	1(0 · 1 + 1)	1
0	1	1	0	0(0 · 1 + 1)	0
0	1	0	1	1(0 · 1 + 0)	0
0	1	0	0	0(0 · 1 + 0)	0
0	0	1	1	1(0 · 0 + 1)	1
0	0	1	0	0(0 · 0 + 1)	0
0	0	0	1	1(0 · 0 + 0)	0
0	0	0	0	0(0 · 0 + 0)	0

Note that in $D \cdot (A \cdot B + C)$ whenever $D = 0$, the result is 0.
Analysis of two types of combination networks reveals significant events.

EXAMPLE 1

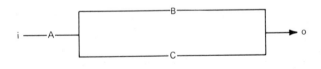

FIGURE 5-16

If A is open, the result will always be 0, regardless of the conditions of B and C. If A is closed, then current is permitted to travel to both the top and lower branches of the parallel network and output then depends upon the conditions of B or C. This event is similar to the effect of switch D in Table 5-8.

EXAMPLE 2

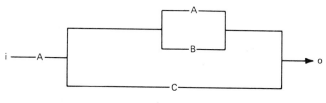

FIGURE 5-17

Assume that two switches have the same variable name A and behave in the same manner. Whenever one A is closed, the other A will also be closed. Whenever one A is open, the other A will also be open. Note that when A is closed, or in the on-condition, B and C have no effect. When A is open, or in the off-condition, B and C cannot have any effect either. The effect is the same as the condition of switch A.

5-7 PROPERTIES OF NETWORKS

Networks have properties similar to the properties of a number system.

In $A + B = B + A$ we express the commutative law for addition, and this law can also be demonstrated in Fig. 5-18. Table 5-9 shows all the possible conditions for the network in Fig. 5-18(a); Table 5-10 shows the same for Fig. 5-18(b).

<table>
<tr><td colspan="4" align="center">Table 5-9</td><td colspan="4" align="center">Table 5-10</td></tr>
<tr><td>A</td><td>B</td><td>A + B</td><td>Result</td><td>B</td><td>A</td><td>B + A</td><td>Result</td></tr>
<tr><td>1</td><td>1</td><td>1 + 1</td><td>1</td><td>1</td><td>1</td><td>1 + 1</td><td>1</td></tr>
<tr><td>1</td><td>0</td><td>1 + 0</td><td>1</td><td>0</td><td>1</td><td>0 + 1</td><td>1</td></tr>
<tr><td>0</td><td>1</td><td>0 + 1</td><td>1</td><td>1</td><td>0</td><td>1 + 0</td><td>1</td></tr>
<tr><td>0</td><td>0</td><td>0 + 0</td><td>0</td><td>0</td><td>0</td><td>0 + 0</td><td>0</td></tr>
</table>

Figure 5-18 and Tables 5-9 and 5-10 show the two networks to be equivalent.

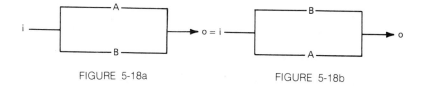

FIGURE 5-18a   FIGURE 5-18b

Definition 5-2

Two networks are said to be *equivalent* if both are open or both are closed for the same state of their corresponding switches.

In Fig. 5-18 we see that when switches A and B are both closed, the result is 1; when either A or B (but not both) is open, the result is 1; and when A and B are both open, the result is 0.

In A · B = B · A we express the commutative law for multiplication. This law can also be demonstrated in equivalent networks and their accompanying tables. See Fig. 5-19 and Tables 5-11 and 5-12.

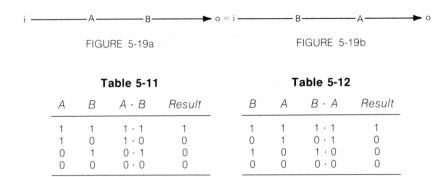

FIGURE 5-19a   FIGURE 5-19b

Table 5-11

A	B	A · B	Result
1	1	1 · 1	1
1	0	1 · 0	0
0	1	0 · 1	0
0	0	0 · 0	0

Table 5-12

B	A	B · A	Result
1	1	1 · 1	1
0	1	0 · 1	0
1	0	1 · 0	0
0	0	0 · 0	0

It is also possible to show equivalences for the circuits that demonstrate the associative law for addition and the associative law for multiplication. These are left as exercises for the student.

Exercises 5-2

1. For the network shown in Fig. 5-20, identify each of the statements that follow as true or false.

   a. Switch X is in parallel with series switches B, C, and D.
   b. Switches G and H are in series with parallel switches E and F.
   c. Switches E and G are in parallel.
   d. Switch A is in series with the rest of the network.

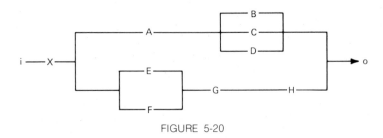

FIGURE 5-20

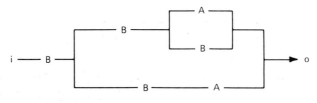

FIGURE 5-21

2. For the network shown in Fig. 5-21
    a. When *B* is closed, can *A* have any effect?
    b. Is output possible when *A* is open?
3. a. Write the equivalent tables illustrating the associative law for addition, where $(A + B) + C = A + (B + C)$.
    b. Write the equivalent tables illustrating the associative law for multiplication where $(A \cdot B) \cdot C = A \cdot (B \cdot C)$.
4. Draw the networks for
    a. $\overline{A}(A + B)$
    b. $AB + CD + EF$
    c. $A(B + BA)$
    d. $X[(A + B)(C + D) + (ABCD)]$
5. Write the tables for a and c in exercise 4.
6. Write the Boolean expressions for

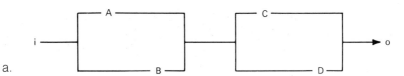

a.

FIGURE 5-22

b.

FIGURE 5-23

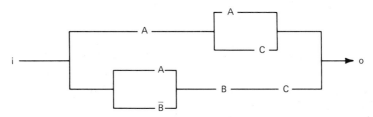

c.

FIGURE 5-24

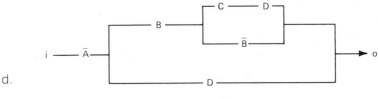

FIGURE 5-25

7. a. Consider a variation on the employment situations in Chapter 4, Exercise 4-2, number 6. We can describe the job prerequisites as:

A = two-year degree in data processing
B = one-year's programming experience
$\overline{B}$ = no programming experience

Draw the appropriate network.

8. Switches have practical applications for programmer control over the computer. The flowchart shown in Fig. 5-26 tests switches and performs several operations, depending upon the condition of the switches. If the switches are off initially, how many records are read? How many lines are printed?

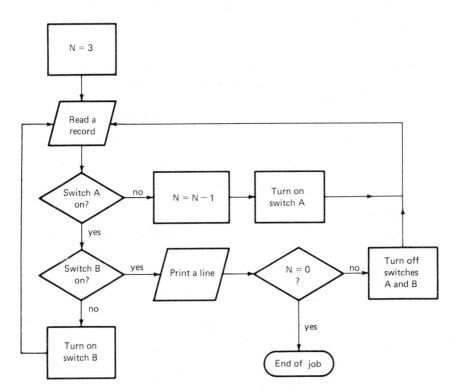

FIGURE 5-26

FIGURE 5-27

5-8 BASIC PROPERTIES OF CIRCUITS

5-8.1 First Distributive Law

In $A \cdot (B + C) = A \cdot B + A \cdot C$ the distributive property exists as it does in conventional algebra. (See the axioms of operations, page 5.) Equivalent networks for this property for each side of the equals sign are shown in Fig. 5-27. Table 5-13 and Table 5-14 show all the possible events for these networks.

Table 5-13					Table 5-14				
A	B	C	$A \cdot (B + C)$	Result	A	B	C	$A \cdot B + A \cdot C$	Result
1	1	1	1(1 + 1)	1	1	1	1	$1 \cdot 1 + 1 \cdot 1$	1
1	1	0	1(1 + 0)	1	1	1	0	$1 \cdot 1 + 1 \cdot 0$	1
1	0	1	1(0 + 1)	1	1	0	1	$1 \cdot 0 + 1 \cdot 1$	1
1	0	0	1(0 + 0)	0	1	0	0	$1 \cdot 0 + 1 \cdot 0$	0
0	1	1	0(1 + 1)	0	0	1	1	$0 \cdot 1 + 0 \cdot 1$	0
0	1	0	0(1 + 0)	0	0	1	0	$0 \cdot 1 + 0 \cdot 0$	0
0	0	1	0(0 + 1)	0	0	0	1	$0 \cdot 0 + 0 \cdot 1$	0
0	0	0	0(0 + 0)	0	0	0	0	$0 \cdot 0 + 0 \cdot 0$	0

Before we can discuss the second distributive property, we must examine a few more properties of switching networks.

5-8.2 Identity Law for Addition

Given: $A + 0 = A$

Since 0 can represent an open switch, we can remove the lower branch in Fig. 5-28, and the output of this network is then entirely dependent upon the condition of A. Thus, 0 is the identity element for addition.

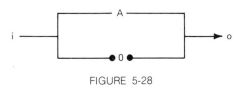

FIGURE 5-28

Consider $A \cdot 0 = 0$

0 can be represented by a broken wire $-A-0\rightarrow$ and there will never be any output in this circuit. In the parallel network the 0 has no effect; in the series network the 0 cancels out the effect of the variable A.

5-8.3 Identity Law for Multiplication

Given $A \cdot 1 = A$

The following networks demonstrate that output depends upon the condition of the variable. The 1 does not represent a switch that can open or close. It means a continuous wire and is written above the line.

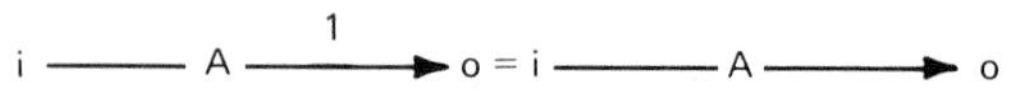

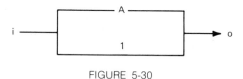

FIGURE 5-29

Table 5-15				Table 5-16	
A	1	A · 1	Result	A	Result
1	1	1 · 1	1	1	1
0	1	0 · 1	0	0	0

The networks shown in Fig. 5-29 are equivalent, as their accompanying tables indicate. Thus, 1 is the identity element for multiplication.

Fig. 5-30 demonstrates that $A + 1 = 1$

```
          A
i  ┌───────────────┐
───┤               ├──→ o
   └───────────────┘
          1
```

FIGURE 5-30

In the parallel network the 1 cancels out the effect of the variable; in the series network the 1 has no effect.

5-8.4 Idempotent Laws

There are two characteristics of two-element Boolean Algebra which have been demonstrated earlier in this chapter. They are:

1. No powers other than the first (or zero) power exist.
2. No coefficients except 1 and 0 are possible.

This means that variables remain unchanged after they have been added to themselves or multiplied by themselves.

For example, in conventional algebra the formula $A + A = 2A$ contains the coefficient 2, where

$$\begin{array}{c} 1 \cdot A \\ \underline{1 \cdot A} \\ 2 \cdot A \end{array}$$

In Boolean Algebra the only coefficients there are are 1 and 0. An inspection of the appropriate network clarifies the expression $A + A = A$, the first idempotent law.

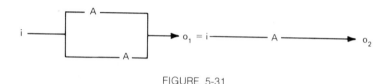

FIGURE 5-31

Since we are only interested in the effect of switches on a network, it is clear that the output at o_1 and o_2 will be dependent only upon the condition of switch A.

In conventional algebra $A \cdot A = A^2$, or a quantity multiplied by itself yields that number to its second power. A characteristic of two-element Boolean Algebra is that all variables exist to the first or zero power and there are no other powers. Inspection of the appropriate network shows $A \cdot A = A$, the second idempotent law.

i —— A —— A ——→ $o_3 = i$ —————— A ————→ o_4

FIGURE 5-32

The output at o_3 and o_4 will be the same and dependent on the on- or off-condition of the switch A.

Table 5-17 summarizes the basic properties of networks. The only property given that we have not discussed is property 14, the Second Distributive Law.

5-8.5 Second Distributive Law

The last property in Table 5-17 is the Second Distributive Law, which we now are able to demonstrate. Using some of the other properties for Boolean networks found in the table, we can show that

$$(A + B) \cdot (A + C) = A + BC$$

Table 5-17 BASIC PROPERTIES OF CIRCUITS

1. $A + B = B + A$
   Commutative law for addition

2. $AB = BA$
   Commutative law for multiplication

3. $(A + B) + C = A + (B + C)$
   Associative law for addition

4. $(AB)C = A(BC)$
   Associative law for multiplication

5. $A(B + C) = AB + AC$
   First distributive law

6. $A \cdot 1 = A$
   Identity law for multiplication

7. $A + 0 = A$
   Identity law for addition

8. $A \cdot 0 = 0$

9. $A + \bar{A} = 1$

10. $A \cdot \bar{A} = 0$

11. $A + 1 = 1$

12. $A + A = A$
    First idempotent law

13. $A \cdot A = A$
    Second idempotent law

14. $A + BC = (A + B)(A + C)$
    Second distributive law

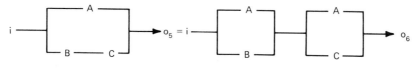

FIGURE 5-33

		Property
to prove:	$A + BC = (A + B)(A + C)$	
step 1	$(A + B)(A + C) = AA + AB + AC + BC$	5
step 2	$= A + AB + AC + BC$	13
step 3	$= A(1 + B + C) + BC$	5
step 4	$= A(1) + BC$	11
step 5	$= A + BC$	6

Note that this is not demonstrable in conventional, non-Boolean algebra (see page 5).

An explanation of the proof is as follows:

Step 1 performs the algebraic multiplication on the given, $(A + B) \cdot (A + C)$.

In Step 2, $A \cdot A = A$ by the second idempotent law.

Step 3 "factors out" the variable A.

In Step 4, $1 + B = 1$ and $1 + C = 1$ by the 11th property.

In Step 5, $A \cdot 1 = A$ by the identity law for multiplication.

Inspection of the networks for the Second Distributive Law reveals that the output of o_5 and o_6 are the same if the conditions of the corresponding variables remain the same. Note that the network on the left in Fig. 5-33 contains fewer switches than its equivalent network on the right.

The network on the left is a simplification of the network on the right.

5-9 SIMPLIFICATION OF CIRCUITS

If there are fewer switches in a network, the saving in production costs is considerable to companies that mass-produce electronic switching networks. Conventional algebra contains many examples of simplification.

EXAMPLE Given:

$$\frac{2n + 6}{n + 3}$$

$$\frac{2n + 6}{n + 3} = \frac{2(n + 3)}{(n + 3)} \qquad \text{distributive law}$$

$$\frac{2(n + 3)}{(n + 3)} = 2 \qquad \text{equals divided by equals are equal}$$

Using the properties in Table 5-17, many different networks can be simplified.

Following are five problems in simplification. Their solutions should be studied carefully.

EXAMPLE 1 Simplify:

	Property
$A + AB$	
$A + AB = A \cdot 1 + AB$	6
$= A(1 + B)$	5
$= A \cdot 1$	11
$A + AB = A$	6

Networks that illustrate this simplification are:

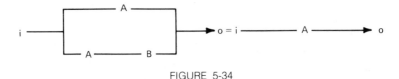

FIGURE 5-34

EXAMPLE 2 Simplify:

$$A + \overline{A}B$$

The second distributive law will help to simplify this network.

	Property
$A + \overline{A}B = (A + \overline{A})(A + B)$	14
$= 1(A + B)$	9
$A + \overline{A}B = A + B$	6

Networks that illustrate this simplification are:

FIGURE 5-35

EXAMPLE 3 Simplify:

	Property
$A + B + \overline{A}$	
$A + B + \overline{A} = A + \overline{A} + B$	1
$= 1 + B$	9
$= 1$	11

Networks that illustrate this simplification are:

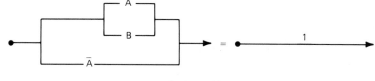

FIGURE 5-36

EXAMPLE 4   Simplify:

                                                              Property

$\overline{A}(B + C) + AB$

$$\overline{A}(B + C) + AB = \overline{A}B + \overline{A}C + AB \qquad\qquad 5$$
$$= \overline{A}B + AB + \overline{A}C \qquad\qquad 1$$
$$= B\overline{A} + BA + \overline{A}C \qquad\qquad 2$$
$$= B(\overline{A} + A) + \overline{A}C \qquad\qquad 5$$
$$= B \cdot 1 + \overline{A}C \qquad\qquad 9$$
$$= B + \overline{A}C \qquad\qquad 6$$

Networks that illustrate this simplification are:

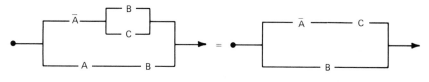

FIGURE 5-37

EXAMPLE 5   Prove:

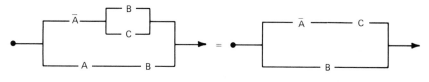

FIGURE 5-38

                                                       Property
$$ABC + A\overline{B}C + \overline{A}\overline{B}C = C(AB + A\overline{B} + \overline{A}\overline{B}) \qquad\qquad \text{2 and a generaliza-}$$
                                                          tion of 5

$$= C[A(B + \overline{B}) + \overline{A}\overline{B}] \qquad\qquad 5$$
$$= C[A(1) + \overline{A}\overline{B}] \qquad\qquad 9$$
$$= C(A + \overline{A}\overline{B}) \qquad\qquad 6$$
$$= C[(A + \overline{A})(A + \overline{B})] \qquad\qquad 14$$
$$= C(1)(A + \overline{B}) \qquad\qquad 9$$
$$= C(A + \overline{B}) \qquad\qquad 6$$

Exercises 5-3

1. Describe the idempotent laws.
2. Give the Boolean expression for:

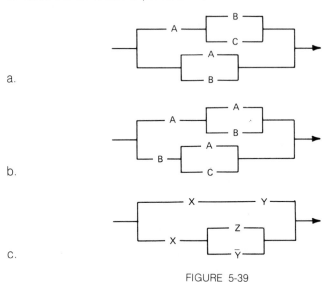

a.

b.

c.

FIGURE 5-39

3. Draw the circuitry and appropriate tables for
   a. $\overline{A}(A + B)$
   b. $\overline{AB}(A + B)$
   c. $AB + AB + A(B + D)$
4. Simplify and list the appropriate property for each step.
   a. $AB + \overline{A}$
   b. $B(AB)$
   c. $A(A + AB)$
   d. $X + \overline{X}Y + Z$
   e. $A + A\overline{B} + A\overline{B}\overline{C}$
   f. $A(B + C) + A + AB$
5. Draw the original circuit and its simplification in 4a, b, c, d, e, and f above.
6. Consider a data processing situation where an application for a credit card will be approved only under the following conditions:
   1. The applicant is over 21 years of age.
   2. The applicant has a good record for paying bills.
   3. The applicant is employed or has a source of income greater than $9,500 a year.
      a. Draw the appropriate network.
      b. What basic property of circuits allows this network to be simplified?

7. An insurance company has male and female policy holders. It wants to search its data base for each of these two groups and determine how many policy holders have over $100,000 insurance with a double indemnity clause for accidental death. Draw the appropriate network and then simplify it, giving the properties that help in simplifying.

5-10 NETWORKS THAT PERFORM A "CARRY"

In Fig. 5-40 let symbol (a) represent a parallel network, symbol (b) represent a series network, and symbol (c) represent a complement. Then the parallel network can be described as an OR GATE, and the series network can be described as an AND GATE. Refer back to the discussions of parallel and series networks to understand why this is so. The complement changes the circuit from the value 1 to 0 or 0 to 1; in other words it inverts the circuit.

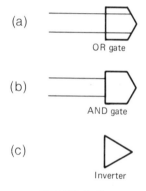

(a) OR gate

(b) AND gate

(c) Inverter

FIGURE 5-40

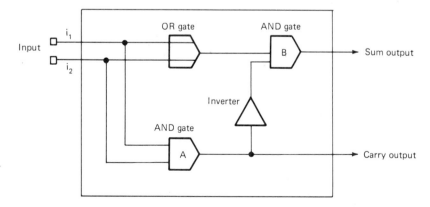

FIGURE 5-41

Figure 5-41 shows a network that is capable of adding two binary digits and producing a carry if necessary. To understand how the network operates, we will consider the simple binary addition

$$
\begin{array}{r}
1 \\
+\ 1 \\
\hline
\text{carry} \quad 10
\end{array}
$$

In Fig. 5-41, the addition of 1 and 1 is represented by two inputs, one at i_1 and one at i_2. Since the OR GATE is a parallel network, the two inputs are passed on as one input to AND GATE B. Further, the two inputs are applied to AND GATE A. Since the AND gate is a series network, the two inputs are passed on as a CARRY OUTPUT. This same output is passed on to the INVERTER. Since the INVERTER passes on an output only if it *does not* receive an input, this means that AND GATE *B* receives only one input and the SUM OUTPUT is zero. Thus, our answer is 10.

Exercises 5-4

1. Trace the circuits in Fig. 5-41 for $1 + 0$, $0 + 1$, $0 + 0$.
2. Two half-adders have been combined in Fig. 5-42 to add three binary digits and provide a sum and carry output. Trace all the possible inputs for i_1, i_2, and i_3, and determine whether *A* should contain:

   a. an AND GATE
   b. an OR GATE
   c. an inverter

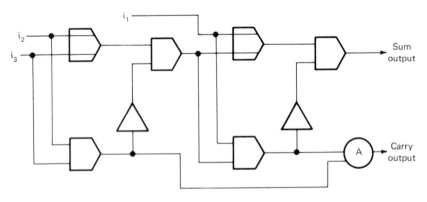

FIGURE 5-42

6

PROGRAMMED
DECISION MAKING

In this chapter we will be applying the material we covered in logic (Chapter 4), numeration systems (Chapter 1), and arithmetic operations (Chapter 2). Computer decision-making will be discussed in some detail.

In order to examine relationships between the binary numeration system and the logic used in networks, it is necessary to define several new terms: *operator, operand, statement label,* and the *IF statement,* which are all found in the *Fortran, Cobol,* PL/I, and Basic programming languages.

6-1 OPERATORS

Definition 6-1

An *operator* is a symbol or a word used to perform a specific computer operation.

In this chapter we are interested in three types of operators: logical operators, relational operators, and arithmetic operators.

I. *Logical* operators

AND
OR
NOT

Relational operators may appear in two different formats, shown separated by commas in the left-hand column below.

II. Relational, or comparison, operators

Operator	Meaning
GE, $>=$	Greater than or equal to
GT, $>$	Greater than
*NE, $\neg=$	Not equal to
EQ, $=$	Equal to
LT, $<$	Less than
LE, $<=$	Less than or equal to
*NL, $\neg<$	Not less than
*NG, $\neg>$	Not greater than

III. Arithmetic operators

Operator	Meaning
$+$	Addition
$-$	Subtraction
$*$	Multiplication
$/$	Division
$**$	Exponentiation

Definition 6-2

An *operand* is a constant or variable that is operated on.

Definition 6-3

A *statement label* is either a number or a name given to a computer statement which contains operators and operands.

All Basic, and many Fortran, statement labels are numbers. Cobol and PL/I statement labels are called names and typically begin with an alphabetic character. For example:

			Operators		
			operand	↓	operand	
Statement	⎰ 135	X =	A	+	B	Fortran or Basic
labels	⎱ HERE	Y =	L	*	3.14	Cobol or PL/I

Statements

The IF statement (with its variations) is used in the Fortran, Basic, Cobol, and *PL/I* computer languages to aid in testing variables and making programmed decisions. A general format for the IF statement is

IF   (condition)   (statement)

---

* The symbol $\neg$ means NOT in *PL/I*.

An example of the use of the IF statement in Cobol helps to illustrate the testing and decision-making procedures:

IF INVENTORY IS LESS THAN
PURCHASE-ORDER GO TO BACK-ORDER.

where BACK-ORDER is a statement label.

The format for the IF statement in Fortran is similar to that of Cobol.

IF(X.LE.51) GO TO 101

is a Fortran statement that says that if X is less than or equal to 51, go to a statement whose label is 101. Other examples of Fortran statements in which the relational operators appear are:

IF (XMAX.GT.10.0) GO TO 98
IF (A.EQ.B) GO TO 75

---------------------------------- **6-2 LOGICAL OPERATORS**

IF statements may contain logical operators. When expressions contain either the AND or the OR operators, the expressions on each side of the operator are evaluated and then compared for their truth values.

The general format is

expression₁   operator   expression₂

For example, if expression₁ were

$$X + Y > A$$

and expression₂ were

$$X/Y = 2$$

we could arrange these two expressions with the AND operator as follows:

$$X + Y > A \text{ AND } X/Y = 2$$

Now if

$$X = 4$$
$$Y = 2$$
$$A = 1$$

then

$$4 + 2 > 1 \text{ AND } 4/2 = 2$$

is true. However X, Y, and A may have other values in a computer program. Thus, there are four possible events in

expression₁   operator   expression₂

These are listed below in Table 6-1.

Table 6-1   Truth Table for "AND"

Truth Value of Expression$_1$	Truth Value of Expression$_2$	Result
1	1	1
1	0	0
0	1	0
0	0	0

Although this is a truth table, we have not used T and F. Instead, a true expression is represented by 1 and a false expression by 0. Note that this truth table is the same as Table 4-1, the truth table for conjunction. (See page 94.)

For the result to be true, i.e., have a value of 1, the expressions on each side of the AND operator must be true. If either or both expressions are false, the result is false, shown by a value of 0.

EXAMPLE 1   Let

$$A = 3, B = 4, C = -5$$

If

$$A + B < C \text{ AND } C < A$$
$$3 + 4 < -5 \text{ AND } -5 < 3$$
$$7 < -5 \text{ AND } -5 < 3$$

We have truth values of   0 AND 1

Thus, result is   0

Hence, since expression$_1$ is false, the result is false.

In a similar manner, Table 6-2 illustrates that there are only four possible events for the OR operator.

Table 6-2   Truth Table for "OR"

Truth Value of Expression$_1$	Truth Value of Expression$_2$	Result
1	1	1
1	0	1
0	1	1
0	0	0

Note that Table 6-2 is the same as Table 4-2, the truth table for disjunction (page 94). For the result to be true, either or both expressions must be true. If both sides are false, the result is false.

EXAMPLE 2 Let

$$I = 7, K = 0, N = 3$$

If

$$I * 3 > N * N \ OR \ N + K = 3$$
$$7 * 3 > 3 * 3 \ OR \ 3 + 0 = 3$$
$$21 > 9 \quad OR \ 3 = 3$$

We have truth values of   1   OR   1

and the result is   1

If we change I to 2 in the example and retain the same values for K and N, the result is still true. However, if $I = 2$, $N = 7$, $K = 3$, the result will be false.

The NOT operator reverses the truth value of the expression it operates on. This is similar to taking the one's complement (page 54) or the action of the inverter in Fig. 5-41.

Using NOT, we can write a series of instructions for an EXCLUSIVE OR situation where our result is true (or 1) if one and only one of the expressions is true (or 1).

EXAMPLE 3 Let $E = (A \ AND \ B) \ OR \ (A \ OR \ B)$ where E will contain the truth value for the entire statement.

First try $A = 0$, $B = 0$.

$$E = (A \ AND \ B) \quad OR \quad (A \ OR \ B)$$
$$0 \ AND \ 0 \quad OR \quad 0 \ OR \ 0$$

The truth values are   0   OR   0

and the result in E is   0

If we try $A = 1$ and $B = 1$, then

$$E = (A \ AND \ B) \quad OR \quad (A \ OR \ B)$$
$$1 \ AND \ 1 \quad OR \quad 1 \ OR \ 1$$

We have truth values of   1   OR   1

and the result in E is   1

But by our definition of EXCLUSIVE OR, only one of the expressions may be true, not both. We need statements that will yield a 0 truth value when both $A = 1$ and $B = 1$. Hence,

Let
$$C = (A \ AND \ B)$$
$$D = (A \ OR \ B)$$

where C will contain the truth value for (A AND B) and D will contain the truth value for (A OR B) in $E = (A \ AND \ B) \ OR \ (A \ OR \ B)$.

Let
$$K = 1$$
$$L = 0$$

Then using the IF statement which tests a condition, we can write

$$\text{IF A} = \text{K GO TO TEST D};$$

XOR: E = C OR D;

—

— other statements in program

—

TEST D: IF (NOT (B = L)) GO TO NEXT;
GO TO XOR;

NEXT: E = L;

(Note: The GO TO causes a branch to the statement whose label follows, i.e., GO TO XOR means that the computer will branch to the statement labeled XOR, namely, E = C OR D, and execute it.)

If A = K i.e., A has a value of 1, and B also has a value of 1, we cannot have the necessary conditions for the EXCLUSIVE OR, and program control is transferred out of the IF statements to the statement labeled NEXT, so that E is set to a truth value of 0. If A or B, but not both, has a truth value of 1, the necessary conditions for the EXCLUSIVE OR will have been met.

As an exercise, write your own EXCLUSIVE OR routine using IF, AND, OR, and NOT with appropriate statements and their accompanying labels.

Exercise 6-1

1. Verify Example 3 by using the four possible combinations of A = 1, A = 0, B = 1, B = 0.
2. Construct the truth table for the EXCLUSIVE OR.
3. Let A = 3, B = −2, C = 4.
   Find the result for:

   a. A − B > C AND C = B ** 2
   b. C * 2 < = 6 OR A * B = A * C
   c. (C + 2)/B.NE.10 OR C.NG.A + A
   d. A + 1 > B + 1 AND C + 1.GT.A

4. Complete:

   a. Fortran statement labels contain _____.
   b. An operand is a constant or _____ that is _____ on.
   c. The IF statement is one of the most powerful computer statements used in a program. Its chief function is _____.
   d. _____ is the exponentiation operator.
   e. NOT is a _____ operator.
   f. = is a _____ operator.

5. Flowchart the EXCLUSIVE OR routine on pages 131–132.
6. Given the following Cobol statements

IF INVENTORY IS LESS THAN PURCHASE-ORDER GO TO TEST-PRICES.

 Other statements in program

TEST-PRICES.
    IF VENDOR-ONE-PRICE IS LESS THAN VENDOR-TWO-PRICE GO TO NEXT-TEST ELSE GO TO READ-NEXT-VENDOR.
NEXT-TEST.
    IF VENDOR-ONE-PRICE IS LESS THAN VENDOR-THREE-PRICE GO TO BACK-ORDER-ONE.

a. What are the only conditions possible to go to the statement BACK-ORDER-ONE ?

b. Describe (in English) how the AND operator can be used to simplify testing for the lowest price.

_____ 6-3 BINARY LOGICAL OPERATORS

The relationship of the binary numeration system to AND and OR instructions on binary operands is easy to follow and has many practical applications.

6-3.1 The AND Operator

When two values are "ANDed" together, their truth values are compared column by column (or position by position).

```
        1 0 1 1 0 0 1 1    Value A
AND     0 1 0 1 0 1 0 1    Value B
        0 0 0 1 0 0 0 1    Result
```

The resulting bit* for each column is 1 if and only if the corresponding bits for value A and value B are both 1; otherwise the resulting bit is 0.

6-3.2 The OR Operator

When two values are "ORed" together, their bits are compared column by column.

```
        0 1 1 0 1 1    Value A
OR      1 0 1 0 0 0    Value B
        1 1 1 0 1 1    Result
```

---

* A bit is a binary integer.

The resulting bit is 1 if either or both corresponding bits from value *A* or value *B* is 1; otherwise the resulting bit is 0.

6-3.3 The EXCLUSIVE OR Operator

When two values are "EXCLUSIVE ORed" together, their bits are compared column by column.

```
                    1 0 1 1      Value A
    EXCLUSIVE OR    0 1 0 1      Value B
                    1 1 1 0      Result
```

The resulting bit is 1 if one and only one of the corresponding bits from value *A* or value *B* is 1; otherwise the resulting bit is 0.

6-3.4 Practical Applications

A practical application of these computer instructions is found in the following problem.

Assume the following descriptions for housing:

Type
- Within city limits
- Public
- Private
- Multiple units

Ordinances
- Screens
- Porch railings
- Water pipes
- Electrical outlets

The first four items describe the type of housing; the last four items refer to city ordinances governing the conditions of all housing.

These descriptions can be expressed in a binary format.

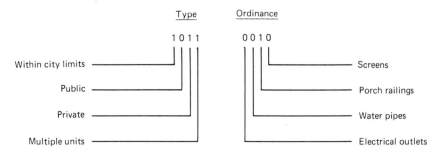

The property in the example above is indicated as an in-city, privately-owned, multiple-unit building with porch railings. Suppose that to meet the requirements of the law, the items under ordinance must all contain ones.

Let value A be the constant for in-city, private multiple units that meet the requirements of the law.

Let value B represent the reference value for any property under consideration. In

```
         10111111    Value A
AND      10110010    Value B
         10110010    Result
```

the result of the AND operation clearly shows that the property under consideration has three violations, or three 0s where there should be 1s.

If we use the EXCLUSIVE OR operation

```
                 10111111    Value A
EXCLUSIVE OR     10110010    Value B
                 00001101    Result
```

the precise violations are indicated as 1s.

The OR operation can also be used to indicate violations.

```
        10111111    Value A
OR      10110010    Value B
        10111111    Result
```

If the owner of value B knows what he has, the result of the OR can reveal what he needs to meet the requirements of the city ordinances.

Exercise 6-2

1. Evaluate for AND.

   a. 10110110   b. 110110   c. 10011111
      01010101   111111   00000000

2. Evaluate for OR.

   a. 10101001   b. 101010   c. 11110101
      00110110   111111   00000000

3. Evaluate for EXCLUSIVE OR.

   a. 10101110   b. 110010   c. 11000101
      10010111   111111   00000000

4. a. Describe the effect of the results in b and c in exercises 1, 2, and 3.
   b. What is the effect of EXCLUSIVE OR on an operand with itself?

5. In each of the following describe the operation as AND, OR, EXCLU-
SIVE OR, addition, or subtraction.

a. 10101010
    11001100
    10001000

b. 10010111
    10001001
    00001110

c. 11101011
    01010111
    10111100

d. 00011010
    01101101
    10000111

e. 110110
    100001
    110111

6. Design an eight-position binary operand that will make decisions for
the following situations:

a. An electronics company is interested in male or female employees
with at least a two-year college degree or four years of experience.
A passing grade on an aptitude test is also required.

b. A special research project requires a graduate student who was on
the dean's list at least in his senior year and has had four years of
mathematics and one year of physics.

_____ 6-4 *AND* AND *OR* OPERATORS IN *PL/I*

The AND operator in *PL/I* is the & symbol.

The OR operator in *PL/I* is the | symbol.

When expressions containing either the AND or the OR operators are
found in *PL/I* statements, the expressions on each side of the operator
are evaluated and then compared for their truth values. The general
format for these operators is

expression₁ OPERATOR expression₂

Four events are possible for both AND and OR in the above format.
Letting 1 represent true and 0 represent false, we can express the events
for AND in tabular form as previously shown in Table 6-1.

Table 6-1 states that if both sides of the statement are true, then the
result is true, and this is indicated by a 1. If either or both sides are false,
the result is false, indicated by 0.

EXAMPLE 1   Let

$$A = 2$$
$$B = 6$$
$$C = 8$$

	AND	
expression₁	*operator*	*expression₂*
A + C > B	&	A * A = (C − B) + A
2 + 8 > 6	&	2 × 2 = (8 − 6) + 2
10 > 6	&	4 = 4
Values 1	&	1
Result	1	

10 is greater than 6 and 4 = 4; therefore, the truth values on each side of the & operator result in 1. When these two values are ANDed together, the result is 1, as indicated by Table 6-1.

EXAMPLE 2   Using the same values for A, B and C as in Example 1, consider

C < A − B	&	C ** 2 ≠ A ** 2 * C
8 < 2 − 6	&	8² ≠ 2² × 8
8 < − 4	&	64 ≠ 32
we have values of 0	&	1
yielding a result of 0		

8 is not less than − 4, results in 0 for expression₁; 64 is not equal to 32, results in 1 for expression₂. When 0 and 1 are ANDed together, the result is 0, as indicated by Table 6-1.

EXAMPLE 3   As an exercise, construct an example where the truth values for expression₁ and expression₂ are both 0, or false.

Table 6-2 states that if either or both sides of a statement is true, then the result is true, indicated by a 1. If both sides of the statement are false, then the result is false, indicated by a 0.

EXAMPLE 4   Now let

$$X = 1$$
$$Y = 3$$
$$Z = 4$$

Then consider

	OR	
expression₁	*operator*	*expression₂*
Z * Y = X * Z × 3	|	Z = X + Y
4 × 3 = 1 × 4 × 3	|	4 = 1 + 3
12 = 12	|	4 = 4
Values 1	|	1
Result	1	

Since both sides of Example 4 are true, the result is true. When the two values in example 4 are ORed together, the result is 1, as indicated by Table 6-2.

EXAMPLE 5 Use the values for X, Y and Z shown in Example 4.

Y + Z < = X + Z	|	Y ** 2 = 2 × Z + 1
3 + 4 < = 1 + 4	|	9 = 2 × 4 + 1
7 < = 5	|	9 = 9
Values 0	|	1
Result	1	

Since one side of Example 5 is true, the result is true by Table 6-2.

EXAMPLE 6 As an exercise, construct an example where the truth values for expression$_1$ and expression$_2$ are both 0 or false.

Consider an operation in a computer program where 1101 is to be ORed with 1011. We could set up the problem as

$$
\begin{array}{r}
1101 \\
\text{OR}\quad \underline{1011} \\
1111
\end{array}
$$

If we as programmers wished to access these resulting *bits,* we could code as the following:

$$
A = 1101
$$
$$
B = 1011
$$
$$
C = A \mid B
$$

The result 1111 is now available to us as the value of C. We can print. store, or do further operations with C.

In performing 1101 & 1011, we could set the problem up as follows:

$$
\begin{array}{r}
1101 \\
\text{AND}\quad \underline{1011} \\
1001
\end{array}
$$

Accessing this result is done using the method in the previous example.

$$
A = 1101
$$
$$
B = 1011
$$
$$
C = A \& B
$$

Then 1001 is available as the value of C.

6-4.1 The Wolowitz Algorithm

An algorithm is a finite set of instructions designed to accomplish a specific task. When programmed for the computer, an algorithm may or may not contain input values, but it should satisfy the following criteria:

1. There is at least one value produced as output.
2. Each instruction is clear and precise.
3. The algorithm terminates after a finite number of instructions.
4. The algorithm performs the required task efficiently and completely.

The Wolowitz Algorithm* states that for a field of n positions or bytes, if the first byte is EXCLUSIVE ORed with the second byte and the result is EXCLUSIVE ORed with the third byte and so on for the length of the field, then the original field will appear after 2^x iterations have been performed (where 2^x is equal to or higher than n).
Consider

Hexadecimal	B C	1 2	4 8
Binary	10111100	00010010	01001000
	10111100	10101110	11100110
	10111100	00010010	11110100
	10111100	10101110	01011010
	10111100	00010010	01001000
Hexadecimal	B C	1 2	4 8

Do the following as an exercise:
Given a hexadecimal field containing D9C9C7C8E3, use the Wolowitz algorithm to find how many iterations are needed to restore the original field. Show all work.

Exercises 6-3

1. Let X = 0, Y = 2, Z = −1. Find the truth value for:
   a. X = Y | Y > Z
   b. Z ** 2 < Y & X + Y + Z = 1
   c. 2 ** Y < X & Y * (Z/2) = −1
   (Show all work.)
2. If X = 1000, Y = 1011, Z = 0000 initially, give the results for:
   a. Z = X & Y
   b. Z = X|Y
3. If X = 101 and Y = 111 and the result of an operation on X and Y is 111, what was the operation?
4. Programmers often do error analysis from printed representation of the contents of internal storage. Given the hexadecimal numbers

	0C	4D	29	3E
Byte numbers	0	1	2	3

found in a printout of internal storage, consider each hexadecimal

---
* Howard J. Wolowitz, The Small Computer Company.

number as a binary quadruplet and give the result in hexadecimal for

a. ORing bytes 0 and 1
b. ANDing bytes 0 and 1
c. EXCLUSIVE ORing bytes 2 and 3 (Note: Each underline indicates a byte.)
d. What is the result of the NOT operator on byte 3?

5. Refer to the housing problem on page 134. Assume that the buildings under consideration were multiple public units and within city limits. Write the hexadecimal constant (as a byte) to identify possible violations.

6-5 COMPARING

Decision making by the computer is frequently performed after the comparison of two values. In an inventory situation we might want to determine if the number of items in inventory equals or is less than a predetermined level number.

Assume that when an inventory contained 100 or fewer stoves, a reorder point was reached and a purchase order had to be sent to a vendor. Further, assume that only one stove is to be sold to a customer at a time. We can flowchart this procedure as shown in Fig. 6-1.

The events in Fig. 6-1 can be described as follows:

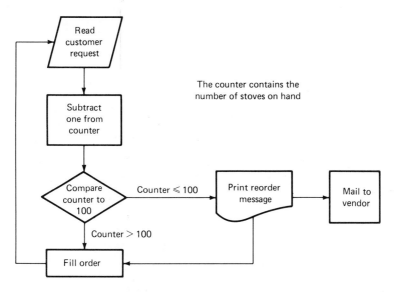

FIGURE 6.1

1. Read the customer's request for one stove.
2. Subtract 1 from COUNTER, where COUNTER is a record containing the number of stoves currently in inventory.
3. Compare COUNTER to 100.
4. If counter is greater than 100, fill the order.
5. If counter is equal to or less than 100, order more stoves, then fill the customer's request.

The last three steps describe a comparison between two values and the actions taken as a result of the comparison.

Calling 100 the constant, and the variable and reference number the COUNTER, we can describe the three possible results of any comparison using Table 6-3 below:

1. If the reference number equals the constant, an equals indicator is turned on.
2. If the reference number is less than 100, a low indicator is turned on.

Table 6-3

Compare		Result
Counter	Constant	
100	100	*Equals* indicator on
99	100	*Low* indicator on
101	100	*High* indicator on

3. If the reference number is greater than 100, a high indicator is turned on.

It is a simple matter for a programmer to test these indicators and write the appropriate instructions or decisions to be made depending upon which indicator had been turned on. Using relational operators and the variable N to represent the current number of stoves, we can write

IF (N.LT.100) GO TO 10
IF (N.EQ.100) GO TO 20
IF (N.GT.100) GO TO 30

where statements numbered 10, 20, and 30 contain appropriate instructions to update the inventory of stoves. As an exercise, write the instruction to GO TO statement number 40 if $N \leq 100$ as shown in Fig. 6-1.

6-6 FORTRAN DECISION MAKING

Another technique used for decision making in Fortran is a test for a positive, negative, or zero result. This early form of the IF statement in

Fortran very clearly demonstrates how an exclusive decision between alternatives is made. In

$$IF \quad (e)\, n_1, n_2, n_3$$

where (e) represents an expression or variable and n_1, n_2, and n_3 are statement numbers, if the value of the expression within the parentheses is negative, then statement n_1 is the next statement to be executed. If the value of the expression is zero, statement n_2 is executed. If the value of the expression is positive, statement n_3 is executed.

Assume that it is necessary to make a choice among three actions depending on the value of a variable X. Suppose further that

If $X < 0$, then statement n_1 is executed.
If $X = 0$, then statement n_2 is executed.
If $X > 0$, then statement n_3 is executed.

The IF statement in Fortran could be written as follows:

$$IF \ (X)\ 9,\ 10,\ 11$$

If we performed an arithmetic operation within the parentheses, we could test for the result in the same manner:

$$IF \ (X - Y)\ 1, 2, 3$$

If X is greater than Y, the computer will go to statement 3; if X equals Y, the computer will go to statement 2; and if X is less than Y, the computer will go to statement 1. These tests are similar to the use of the relational operators. In

$$IF \ (X - Y)\ 1, 2, 2$$

we are effectively saying

$$IF \ (X.GE.Y) \ GO \ TO \ 2$$

Consider another example. Assume that distinct values have been computed for three variables X, Y, and Z. If the value of X is

a. greater than Y and greater than Z, go to statement number 10.
b. less than Y but greater than Z, go to statement number 20.
c. greater than Y but less than Z, go to statement number 30.
d. less than Y and less than Z, go to statement number 40.

Write a sequence of IF statements to make the decisions outlined above. (The second statement number option is irrelevant in each IF statement, since it is given that no two values are alike.) One solution is:

$$IF \ (X - Z)\ 70, 50, 60$$
$$60 \quad IF \ (X - Y)\ 20, 50, 10$$
$$70 \quad IF \ (X - Y)\ 40, 50, 30$$

Consider next a data processing example in Fortran that illustrates the use of the IF statement in decision making. A car dealer makes an agreement with each of his salesmen to give them

1. a guaranteed salary of $1000 per month
2. 5% commission on all sales
3. 10% bonus on all sales over $20,000.

Assume that at least one car a month is sold, and use the IF statement to test for sales, as follows:

```
BONUS = 0.0
COMM = SALES * 0.05
IF(SALES .LE. 20000) GO TO 5
BONUS = (SALES - 20000) * 0.1
5 TOTAL = SALARY + COMM + BONUS
```

Note that we set the variable BONUS to 0.0 initially. If $20,000 in sales has not been achieved, our program will branch around the calculations for BONUS. If the conditions of the IF statement are not met, our program will "fall through" to the next sequential statement and calculate the BONUS.

Exercises 6-4

1. Assume that in a department store, when a customer's charge account has $0.00 balance, no bill is sent to the customer. When the balance due is positive, i.e., the customer owes the store money for purchases made, a bill is sent to the customer. When the balance due is negative, i.e., the customer has either returned items to the store or overpaid, a check for this negative amount is sent to the customer. Construct a flowchart to illustrate these procedures and write the appropriate Fortran coding.

2. Given X = 3, Y = 2, Z = -1. To which statement will the program branch?

   a. IF (X - Y) 1, 2, 3

   b. Y = Y * Z + Z (Y is replaced by the value of the expression on the right of the equals sign, and Y * Z is performed first.)
      IF (X + Y) 1, 2, 3

   c. X = X * Y (X is replaced by itself times Y)
      IF (Z - X) 1, 2, 3

   d. Y = -Y
      IF (X - Y) 1, 2, 3

3. Using the format of the Fortran IF statement found in Exercise 2, write a Fortran statement for:

   a. if $A \geq B$, go to statement 10; otherwise go to statement 20.
   b. if $A \leq B$, go to statement 10; otherwise go to statement 20.

4. When we want to increment a variable in Fortran, we may write X = X + 1, where X equals itself plus one.

Given the following program segment:

$$X = A$$
$$Y = 1$$
$$10 \quad A = A - B$$
$$X = X + A$$
$$Y = Y + 1$$
$$IF \ (Y - Z) \ 10, \ 20, \ 20$$
$$20 \quad X = Y$$

a. If Z has a value of 15 initially, statement 10 will be executed _____ times.

b. If Z has a value of 15 initially, X will have the value _____ after statement 20 is executed.

5. Complete the Fortran statements that follow to calculate gross pay. OVRTM represents overtime and is computed as 1.5 times the number of hours greater than 40.

```
OVRTM = 0.0
IF (HOURS _____ ) _____
10 _____
5 _____ = RATE * HOURS + OVRTM
```

7

SCIENTIFIC NOTATION

7-1 **BASIC CONCEPTS**

Position value and place value provide the decimal numeration system with a compact method of writing very large and very small numbers. This method is called *scientific notation.* Many college mathematical and scientific computer programs can be handled by this method.

Problems in engineering, physics, chemistry, biology, and astronomy require that quantities must be expressed in a manner that is easily read and understood. For example, the speed of light

186,000 miles per second

can also be written as

$$186,000 = 1.86 \times 100,000 = 1.86 \times 10^5$$

Other examples are:

1. 3.16×10^7 seconds in a year
2. 1.44×10^3 minutes in a day
3. 2.09×10^7 feet, radius of the earth
4. 3.83×10^{22} cubic feet, volume of the earth
5. 1.32×10^{25} pounds, mass of the earth
6. 3.69×10^{-27} pounds, mass of a hydrogen atom
7. 2.01×10^{-30} pounds, mass of an electron

In scientific notation it is conventional to express numbers with one digit to the left of the decimal point. Then the resulting number is multiplied by the proper power of ten to yield the correct value. The following examples illustrate this practice:

$$234,000 = 2.34 \times 10^5$$
$$0.234 \times 10^6 = 2.34 \times 10^5$$
$$0.567 = 5.67 \times 10^{-1}$$
$$0.0234 = 2.34 \times 10^{-2}$$
$$0.0789 \times 10^3 = 7.89 \times 10^{-2} \times 10^3$$
$$= 7.89 \times 10$$
$$0.0000567 = 5.67 \times 10^{-5}$$

Performing arithmetic operations on numbers expressed in scientific notation requires care in observing the rules for exponents and decimal point alignment. (Refer to appendix B if necessary.) In

$$0.00023 \times 186,000 = (2.30 \times 10^{-4}) \times (1.86 \times 10^5)$$

we can use the commutative law for multiplication to make possible a more convenient grouping, thus:

$$(2.30 \times 1.86) \times (10^{-4} \times 10^5)$$

Applying the laws of exponents and multiplication, we get

$$(4.278 \times 10) = 42.78$$

Following are two more examples that illustrate the point:

EXAMPLE 1

$$\frac{136,000 \times 4,500,000}{7,800,000} = \frac{(1.36 \times 10^5) \times (4.5 \times 10^6)}{(7.8 \times 10^6)}$$

$$= \frac{(1.36 \times 4.5) \times (10^5 \times 10^6 \times 10^{-6})}{7.8} = \frac{6.12 \times 100,000}{7.8} = 78461.54$$

EXAMPLE 2

$$\frac{456,000 \times 0.00005}{0.0034 \times 19,000,000} = \frac{(4.56 \times 10^5) \times (5 \times 10^{-5})}{(3.4 \times 10^{-3}) \times (1.9 \times 10^7)}$$

$$= \frac{(4.56 \times 5) \times (10^5 \times 10^{-5})}{(3.4 \times 1.9) \times (10^{-3} \times 10^7)} = \frac{22.8}{6.46 \times 10,000} = .0004$$

Exercises 7-1

Express Exercises 1 tnrough 20 in scientific notation.

1. 12345
2. 73900
3. 13.76
4. 14.92
5. 1789.2

6. 381
7. 381,000,000
8. 92,000
9. 746.2×10^{-3}
10. 14.9×10^2
11. 0.0012
12. 0.123
13. 0.00000678
14. 0.0148×10^{-2}
15. 0.0456×10^3
16. 0.951×10^{-2}
17. 0.2126×10^{-4}
18. 0.6×10^2
19. 0.1401×10^{-3}
20. 0.00000123×10^7

In Exercises 21 through 29, perform the indicated operation and write the result in scientific notation.

21. $5,000 \times 0.42$
22. $1,000,000,000 \times 789,000$
23. $4,560,000 \times 0.000021$
24. 0.0987×0.00000672
25. $0.00321 \times 890,000$
26. $0.00642 \times 0.000000789$
27. $\dfrac{0.0031}{4,276}$
28. $\dfrac{7,890,000 \times 4.302}{876,000}$
29. $\dfrac{0.3214 \times 0.0004}{0.76}$

7-2 SIGNIFICANT DIGITS

In the Introduction we noted that *rational* numbers can be put in the form a/b, where $a \in J$ and $b \in J$ for $b \neq 0$. Some rational numbers, e.g., $\frac{1}{3}$ or $\frac{2}{7}$, form a repeating decimal pattern.

$$\frac{1}{3} = .3333 \ldots$$
$$\frac{2}{7} = .285714285714. \ldots$$

Irrational numbers cannot be put in the form a/b. Among such numbers are values for $\sqrt{2}$ and π, which may be approximated to hundreds of decimal positions in which no repetitive pattern exists.

In computer programming it is frequently important to make approximations for irrational and rational numbers, limiting the numbers of digits to a predetermined size. A data item may be limited to six decimal positions in the interest of storage space, processing time, or readability. In this situation ⅓ would appear as .333333, ⅔ as .285714, and π as 3.14159. The *approximation* of ⅔ to six decimal positions contains six *significant digits.*

The following five statements describe the notion of significant digits:

1. All nonzero digits are significant. For example, 134, 2.467, and .85 contain three, four, and two significant digits, respectively.
2. All zeros occurring *between* nonzero digits are significant digits. For example, 10203 is a number with five significant digits; 40.123 is a number with five significant digits.
3. Trailing zeros following a decimal point are significant digits. For example, 45.60, 4.560, .4560, and 56.00 have four significant digits each.
4. Zeros between the decimal point and preceding a nonzero digit are *not* significant digits. For example,

    0.078, which can be expressed as 78 × 10⁻³ contains two significant digits;
    0.000012 contains two significant digits;
    0.012345 contains five significant digits;
    0.00400 contains three significant digits.
5. When the decimal point is not written, trailing zeros are not considered to be significant. 3,000 may be written as 3 × 10³ and contains only one significant digit; 4,560,000 may be written as 456 × 10⁴ and contains three significant digits. 4,560,000.00, however, contains a decimal point and nine significant digits.

   In discussing significant digits and decimal points we are concerned with *accuracy* and *precision.*

Definition 7-1

Accuracy refers to the number of significant digits. π = 3.14159 is accurate to six significant digits. $\sqrt{2}$ = 1.4142 is accurate to five significant digits.

Definition 7-2

Precision refers to the number of decimal positions. ⅔ = .28571428571 has a greater precision than ⅔ = .285714286.

The numbers 123.45, 1.23 and 0.12 are all expressed to two decimal places and have the *same* precision. However, they do not contain the same number of significant digits and therefore differ in accuracy.

Exercises 7-2

1. Which of the following has the greatest precision?

   a. 1.2304   b. 1.23   c. 1.2304064

2. What is the *accuracy* of each of the following?

   a. 41   g. 3000.1
   b. 40.1   h. 0.01
   c. 41.00   i. 42,360.00
   d. 42,360,000   j. 40
   e. .0012   k. 10.000
   f. .00120   l. .10

3. Use scientific notation to set up the solutions to the following problems.

   a. If the average distance from the earth to its moon is about 240,000 miles, how long would it take to travel this distance at 55 miles per hour? Express the answer with an accuracy of six significant digits.

   b. If it took a service bureau one-fourth of a year to complete a payroll system of programs, for how many minutes of time would the programmer charge? Use an eight-hour day, forty-hour week, and express the answer with a precision of five digits.

   c. How fast is the earth moving in its orbit around the sun? Omit leap-year calculations.

7-3 TRUNCATING

   Consider a situation in which a programmer does not want to *round off* numbers. Assume that the operation $0.256 \times 1.6 = 0.4096$ is done on a computer that prints out the answer to an accuracy of three significant digits. The result would be .409, with the low-order 6 truncated or lost. In many computer programs, fifteen digits is all that is permitted. In this situation all digits in excess of fifteen are dropped. Often, the programmer may first determine the size of his numeric data items. Then, in the case of a number such as .285714285714, he may code for as few significant digits as he requires:

Data Item	Significant Digits
.285714285714	12
.28571428571	11
.2857142857	10
.285714285	9
.	.
.	.
.	.
2	1

We noted that 3000 contained only one significant digit, since 3 × 10^3 = 3000. For this reason, numbers such as 3456.78 can also be reduced by progressive truncation to one significant digit in the following manner.

Data Item	Significant Digits
3456.78	6
3456.7	5
3456	4
3450	3
3400	2
3000	1

7-4 ROUNDING

Consider the statements

6.24 is approximately 6.2.
9.99 is approximately 10.
24.6005 is approximately 24.601.

Note that if the digit to be dropped is less than 5, the number is truncated; otherwise the next digit to its left is increased by 1.

The technique of *half-adjusting* is used to round to the least significant digit to be kept. (See also page 285.)

EXAMPLE 1:

$$
\begin{array}{r}
2.407 \\
+ \quad .005 \leftarrow \text{number to half-adjust} \\
\hline
2.41|2 \leftarrow \text{digit to be dropped}
\end{array}
$$

A 5 is added to the digit to be dropped. If the digit to be dropped is equal to or greater than 5, rounding, or a carry of 1, is forced on the next digit to the left.

Truncation does not always provide as desirable a result as *rounding off*, especially in such monetary problems as payroll, billing charge accounts, and preparing budgets. In the number 7.3, for example, .3 is nearer 7 than 8. In the number 7.7, .7 is nearer 8 than 7. We could say that 7.3 is approximately 7 and 7.7 is approximately 8. In saying 7.3 is approximately 7, we are effectively truncating the .3. In saying 7.7 is approximately 8, we are *rounding* to 8, or increasing 7 by 1.

EXAMPLE 2:

$$
\begin{array}{r}
1.09 \\
+ \quad .05 \leftarrow \text{number to half-adjust} \\
\hline
1.1|4 \leftarrow \text{digit to be dropped}
\end{array}
$$

EXAMPLE 3:

$$
\begin{array}{r}
24.971 \\
+ \quad .005 \\
\hline
24.97\boxed{6} \leftarrow \text{digit to be dropped}
\end{array}
$$

EXAMPLE 4:

$$
\begin{array}{r}
42.803691 \\
+ \quad .500000 \\
\hline
43.\boxed{303691} \leftarrow \text{digits to be dropped}
\end{array}
$$

EXAMPLE 5:

$$
\begin{array}{r}
7.1243 \\
+ \quad .0005 \\
\hline
7.124\boxed{8} \leftarrow \text{digit to be dropped}
\end{array}
$$

EXAMPLE 6:

$$
\begin{array}{r}
9.8104 \\
+ \quad .0500 \\
\hline
9.8\boxed{604} \leftarrow \text{digits to be dropped}
\end{array}
$$

Exercises 7-3

1. Truncate to three significant digits:
   a. 123456
   b. 12700000
   c. 12.3456
   d. 43.01
2. Truncate to four significant digits:
   a. 0.00045678
   b. 0.1234
   c. 679.0
   d. 0.14789
3. Use half-adjusting to round off each of the numbers in Exercise 1 to two significant digits.
4. Use half-adjusting to round off each of the numbers in Exercise 2 to three significant digits.

7-5 FIXED-POINT AND FLOATING-POINT NUMBERS IN *PL/I*

In the *PL/I* computer language, most of the arithmetic data operated on are real numbers. Real numbers include fixed-point and floating-point numbers. (Fortran fixed-point and floating-point numbers are discussed in Section 7-7.)

A *fixed-point* number is a real number and may be an integer such as 7 or a number expressed in fractional form such as 43.72.

A *floating-point* number is a real number expressed in terms similar to scientific notation, in which a number is expressed as a fraction between 0.1 and 1.0 times a power of 10.

Decimal floating-point numbers are indicated by a group of decimal digits followed by an E. The decimal number may be signed (plus or minus) or unsigned and may or may not include a decimal point. Digits to the right of the E must not have an included decimal point but may also be signed or unsigned.

To convert a decimal floating-point constant to its decimal fixed-point equivalent, multiply the digits to the left of the E by 10 raised to the power indicated by the number to the right of the E. Study the following examples.

Floating-Point Number	Conversion		Fixed Point Number
1.23E6	1.23×10^6	=	1230000.
71.3E+4	71.3×10^4	=	713000.
4.1E−3	4.1×10^{-3}	=	.0041
8610.2E+2	8610.2×10^2	=	861020.
−41E5	-41×10^5	=	−4100000.

7-6 FIXED-POINT AND FLOATING-POINT NUMBERS IN BINARY

PL/I binary constants contain a B annexed to the low-order position: 1011B is a binary fixed-point constant.

−101.1B is a binary fixed-point constant with one binary fractional position.

Binary floating-point numbers are indicated by a group of binary digits followed by an E. The numbers to the left of the E are binary floating-point numbers which may be signed or unsigned and may or may not include a binary point.

Digits to the right of the E are written as a decimal number. These numbers must not have an included decimal point but may also be signed or unsigned. They are always followed by a B.

111E6B is a binary floating-point constant.

10101E−7B is a binary floating-point constant.

To convert a binary floating-point constant to its binary fixed-point equivalent, multiply the digits to the left of the E by $10_{(B)}$ raised to the power indicated by the number to the right of the E. Recall that 10 means two in the binary numeration system.

Floating-Point Number	Conversion	Fixed-Point Number
1.101E4B	1.101×10^4	11010B
111E3B	111×10^3	111000B
11.10E − 3B	11.10×10^{-3}	.01110B
−1.10E2B	-1.10×10^2	−110.0B

Exercises 7-4

1. a. The letter E is found in _____ numbers.
   b. The letter B is always annexed to the _____-order position of _____ numbers.

2. Change the following to fixed-point numbers.
   a. 1.34E3
   b. 2.31E − 5
   c. 278.1E − 2
   d. − 4.3E4
   e. 1.001E3B
   f. 1.1E − 2B
   g. 1.24E − 4

3. Sections 7-5 and 7-6 imply that in PL/I all _____ data is found in the form of binary or decimal numbers.

7-7 FIXED-POINT AND
_____ FLOATING-POINT NUMBERS IN FORTRAN

The Fortran computer language provides two different kinds of arithmetic: floating-point and fixed-point, or integer, arithmetic. (Note the differences in the meanings of these terms for Fortran in the text that follows.)

The *fixed-point* mode of arithmetic uses integers, or whole numbers, and a Fortran integer is always an *integer* in the arithmetic sense. (See page 2.)

The floating-point mode of arithmetic uses *real* numbers, and the set of real numbers includes integers as well as those numbers which have a decimal fraction part.

Fortran executes computations in such a way that the programmer does not have to be concerned about the location of the decimal point.

Examples of fixed-point and floating-point numbers are:

Fixed-Point	Floating-Point
4	3.14159
27	25.267
64238	− 3.00001
− 120	+ 1.
+ 71	0.0000000001

In 64238 we note that no commas are used in Fortran numbers.* Further, because it contains the decimal point, +1. will be recognized by the computer as a floating-point number. The decimal point is a convenient way to distinguish between the two modes of arithmetic.

In scientific notation, we write the number 7,300,000 as 7.3×10^6. In Fortran we can write this term as $7.3E + 6$. The letter E must be written directly after the constant to indicate a floating-point constant *times* a power of ten. The power of ten may be preceded by a plus sign or a minus sign located immediately to the right of the E. (The plus sign may be omitted.)

For example, $2.1E - 7$ means $2.1 \times 10^{-7} = 0.00000021$, and $4.3E2$ means $4.3 \times 10^2 = 430$.

7-7.1   A Note on Fixed-Point Arithmetic

The fixed-point arithmetic mode permits only integer results. A problem such as $\frac{7}{2}$ results in a quotient of 3; the fraction $\frac{1}{2}$ is truncated or dropped. In many engineering calculations where perhaps seven or eight significant digits are used, only an approximation is necessary. If the programmer attempts to process more than the maximum number of significant digits, the excess digits will be truncated. (Examples of truncation and rounding for business data processing applications are given in Chapter 13.)

The hierarchy, or order, of arithmetic operations in most computer languages is:

Operator or Symbol	Operation	Level
**	exponentiation	1
*, /	multiplication, division	2
+, −	addition, subtraction	3

In any arithmetic expression exponentiation will be performed first, multiplication and division will be performed next, and addition and subtraction last.

EXAMPLE   If

$$J = 3, K = 4, L = -2, M = 0$$

substituting in the expression

$$J * K + L + M$$

---

* Commas are not computational in any computer language.

yields

$$(3 \times 4) + (-2) + 0 = 12 + (-2) = 10$$

Exercises 7-5

Use the values of J, K, L, and M given in the example when you work the following Fortran problems.

1. Find the values for the following in the fixed-point mode:†
   a. J + M
   b. J ** 2 + L * J
   c. K/L + J ** L
2. If M = J + K means that M is *replaced* by J + K, find values for M and J in
   a. M = K * L + M + K ** 2
   b. J = J + 1
   c. J = −J
3. Let A = J, B = K, C = L, and D = M. In the following problems begin your computation with the innermost parentheses as does the computer. Compute the value of each expression in the floating-point mode.
   a. A + (A * B)
   b. (A * (A + B)) + C
   c. (A + (A * B))/C
4. Use the values given in Problem 3 for A, B, C, and D and the hierarchy of operations. If operations at the same level are computed from left to right, find X in
   a. X = B + C / A * B
   b. X = (B + C) / A * B
   c. X = D − A * B + C / B
5. If $A = 1.23 \times 10^{-3}$ and $B = 7.85 \times 10^2$, what is the result of
   a. A + B
   b. A = B + (−B)
   c. A * B

7-7.2 Notes on Handling Fixed-Point and Floating-Point Numbers in Basic

Systems written in the Basic programming language vary, but many provide six to nine significant digits for numeric output. Basic handles

---

† Fixed-point variable names in Fortran begin with one of the following letters: I, J, K, L, M, or N. All other letters are used as the first letter for floating-point variables.

integers and numbers containing decimal fractions by keeping track of the decimal point. This is similar to Fortran systems. For example,

$$100 \quad N = 34.2$$
$$110 \quad L = N * 2$$
$$120 \quad PRINT \ L$$

results in 68.4 for the variable L. (Note that every statement in Basic must have a number and these numbers must be in ascending sequence).

Care must be taken when using decimal points in scientific notation. On some systems the following events are possible.

$$100 \ PRINT \quad 4.2E-1 + 4.2E-1$$

yields a result of .84, which may not be what the programmer wants. However, as another example,

$$100 \ PRINT \quad 10E+10 * 1000E+20$$

yields a result of $1E+34$. In this case, the operands and the results are in the same notation, or the E-format (exponential format).

To force the result into E-format when it is desirable to do so, programmers must write field specifiers. Consider the following program:

$$100 \ N = 54.2$$
$$110 \ T\$ = "\#\#\#.\#\#\# \uparrow \uparrow \uparrow \uparrow "$$
$$120 \ N = N * 10E+1$$
$$130 \ PRINT \ USING \ T\$;N$$

OUTPUT $54.200E+02$

(Note: Not all Basic compilers support USING.)

In statement 110 we define field specifiers in T$, "forcing" the system to print the result in E-format. Field specifiers are contained in quotes and assigned to a variable with an equals sign.

The # sign is a placeholder for any digit to the right or left of the decimal point. In other words, # signs will be replaced by digits in the result. The position of the decimal point must be specified by the programmer.

The ↑ symbols specify that the E-format is to be used. The number of ↑ symbols indicates the number of positions that are required to print the exponential part of the result. One caution is always to be noted: When writing Basic programs that require computations, the programmer should reference the manual for the computer being used. On some Basic systems, for example, the double asterisk, **, specifies exponentiation (as in Fortran and PL/I); on others exponentiation is specified as an upwards-pointing arrow, ↑ .

_____ **7-8 CONVERSIONS OF FRACTIONS**

Using the base or the powers of the base of a number, it is possible to convert decimal fractions to binary, octal, or hexadecimal fractions and vice versa. We will demonstrate the algorithms or methods used to make these conversions, but the mathematical proofs will not be discussed.

7-8.1 Decimal to Binary

Decimal fractions may be converted to binary fractions using the multiplier 2 in a series of multiplications. Each decimal integer 1 or 0 that results from these multiplications forms the binary fraction. Multiplication is continued until the decimal fraction has been reduced to zero. An exception to this is the case of nonterminating fractions where a significant number of binary digits must be generated.

EXAMPLE   Convert the decimal fraction 0.8125 into a binary fraction.

Decimal Fraction	Multiplier	Product	Binary Digits
.8125	× 2	= 1.6250	1
.6250	× 2	= 1.2500	1
.2500	× 2	= 0.5000	0
.5000	× 2	= 1.0000	1

Multiplying .8125 by 2 yields 1.6250. The decimal integer 1 becomes the first binary digit. The second multiplication uses the resulting decimal fraction, .6250, as its multiplicand. After the fourth multiplication, the resulting decimal fraction is .0000 and we need not multiply any more. Digits in the last column on the right are dropped to the left.

Thus, $0.8125_{(D)} = 0.1101_{(B)}$.

7-8.2 Decimal to Octal

Using 8 as a multiplier, we can convert decimal fractions to octal fractions.

EXAMPLE 1   Converting the decimal fraction 0.975 to an octal fraction.

Decimal Fraction	Multiplier	Product	Octal Digits
.975	× 8	= 7.800	7
.8	× 8	= 6.4	6
.4	× 8	= 3.2	3
.2	× 8	= 1.6	1
.6	× 8	= 4.8	4
.8	× 8	= 6.4	6
.4	× 8	= 3.2	3
.2	× 8	= 1.6	1
.6	× 8	= 4.8	4

Dropping the octal digits to the left, we obtain the repeating octal fraction .763146314. . . . Nine digits are shown in this example since a sufficient number of significant digits was generated to illustrate the algorithm. Converting $0.975_{(D)}$ to octal 0.763146314 . . . provides an example of a result that is a nonterminating fraction.

7-8.3   Decimal to Hexadecimal

Using 16 as a multiplier, we can convert decimal fractions to hexadecimal fractions. (Decimal numbers 10 through 15 become hexadecimal digits A through F: see page 35.)

EXAMPLE 2   Convert the decimal fraction 0.69921875 to a hexadecimal fraction.

Decimal Fraction	Multiplier	Product	Hexadecimal Digits
.69921875	× 16	= 11.18750000	B
.1875	× 16	= 3.0000	3

The most significant hexadecimal digit B is placed immediately to the right of the hexadecimal point, and the 3 is placed after the B. Thus, $0.B3$ is our resulting hexadecimal fraction.

7-8.4   Binary to Decimal

Converting binary fractions to decimal requires division.

EXAMPLE 3   Convert .1011 to a decimal fraction.
We express each digit in the appropriate decimal equivalent of the powers of base two thus:

$$.1011 = (1 \times 2^{-1}) + (0 \times 2^{-2}) + (1 \times 2^{-3}) + (1 \times 2^{-4})$$
$$= \tfrac{1}{2} + 0 + \tfrac{1}{8} + \tfrac{1}{16} \text{ (expressed as fractions)}$$
$$= \tfrac{11}{16} \text{ (adding up the terms)}$$
$$= .6875 \text{ (dividing)}$$

7-8.5   Octal to Decimal

The procedure for converting octal fractions to decimal fractions is similar to the one already described for binary fractions.

EXAMPLE 4 $.204_{(O)} = (2 \times 8^{-1}) + (0 \times 8^{-2}) + (4 \times 8^{-3})$
$$= \tfrac{2}{8} + 0 + \tfrac{4}{512} \text{ (expressed as fractions)}$$
$$= \tfrac{33}{128} \text{ (adding up the terms)}$$
$$= .2578125 \text{ (dividing)}$$

7-8.6   Hexadecimal to Decimal

Converting hexadecimal fractions to decimal fractions can be performed using repeated divisions by the decimal number 16.

EXAMPLE 5   Convert .F8 to a decimal fraction.

Begin by dividing the decimal equivalent of the low-order, or right-most, hexadecimal digit by 16:

$$8 \div 16 = .5$$

Add in the decimal equivalent of the next digit to the left (F = 15):

$$.5 + 15 = 15.5$$

Divide this number by 16: $15.5 \div 16 = .96875$

7-9   RELATIONSHIP OF
_____ HEXADECIMAL TO BINARY FRACTIONS

We are able to write a hexadecimal number by marking off a group of binary digits into quadruplets, annexing zeros to the left if needed to complete a quadruplet on the left of the binary point. This method was discussed in Chapter 1.

EXAMPLE 6 $0010 \quad 1101_{(B)} = 2D_{(H)}$

From our last example, we know that the hexadecimal fraction $.F8_{(H)} = .96875_{(D)}$. Now, to express .F8 as a binary fraction, we annex zeros to the *right* to provide a full quadruplet, thus:

$$.1111 \quad 1000 \quad \text{in binary}$$
means   .F8   in hexadecimal

To determine that $.1111 \quad 1000_{(B)}$ is also equal to $.96875_{(D)}$, we need only use the algorithm for converting binary fractions to decimal fractions. Thus:

$$\tfrac{1}{2} + \tfrac{1}{4} + \tfrac{1}{8} + \tfrac{1}{16} + \tfrac{1}{32} = .96875$$

7-10 RELATIONSHIP
OF OCTAL TO BINARY FRACTIONS

In a similar manner, we can convert binary fractions to octal fractions by grouping binary triplets starting from the binary point.

EXAMPLE 7   .101,001   in binary

      means   .51   in octal

Exercise 7-6

Complete the following table:

Decimal	Binary	Octal	Hexadecimal
0.92			
	1101111.10011000		
			1.00C
		276.532	

8

RELATIONS AND FUNCTIONS IN FIRST DEGREE EQUATIONS

In programming computers it is often necessary to translate a problem statement into a mathematical equation. For example, the interest on a short-term business loan can be expressed in the following equation:

INTEREST = (AMOUNT × DAYS × RATE) ÷ 360

Our study of equations begins with *ordered pairs*.

8-1 ORDERED PAIRS

Up to now the order of elements in a set has had no effect on our definitions or operations. In operations on ordered pairs, however, the order of elements in a set is of importance.

Definition 8-1

An ordered pair (x, y) is formed by taking x from a set and designating it as the *first* element and taking y from a set and designating it as the *second* element. x and y may come from different sets or from the same set.

An ordered pair can be represented by a point on a graph. First, to construct the two axes of a graph, draw two perpendicular lines. The

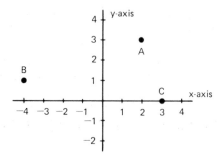

FIGURE 8-1

point at which these axes intersect is called the origin. The horizontal line is usually called the x-axis. The vertical line is usually called the y-axis.

To locate points on a graph, we use a pair of numbers called the x- and y-coordinates of the point. The x-coordinate is sometimes called the abscissa; the y-coordinate is sometimes called the ordinate. The plane on which we draw these points is called the Cartesian coordinate system.

In Fig. 8-1, A has the coordinates (2, 3), B has the coordinates (−4, 1), and C has the coordinates (3, 0). In the parentheses the x, or horizontal distance, is always given first, and the y, or the vertical distance, is given second. Any point on the graph is represented by one unique ordered pair. The rectangular lattice of points in Fig. 8-2 is formed from the set of points (or coordinates) {(1, 1), (1, 2), (1, 3), (2, 1), (2, 2), (2, 3), (3, 1), (3, 2), (3, 3)}. The origin has the coordinates (0, 0).

We note that two ordered pairs (x, y) and (a, b) are equal if and only if x = a and y = b.

It is clear that (3, 2) is not the same ordered pair as (2, 3). However, if a = 3 and x = 3 and b = 2 and y = 2, (x, y) is the same as (a, b) and is not equal to any other ordered pair.

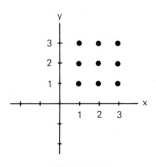

FIGURE 8-2

Definition 8-2

A Cartesian product, designated by $A \times B$, of sets A and B, is the set of all the ordered pairs (x, y) where $x \in A$ and $y \in B$. (Thus, the first element is taken from set A; the second from set B.)

If

$$A = 1, 2, 3, 4$$
$$B = 5, 6, 7, 8$$

then

$$A \times B = \begin{bmatrix} (1,\ 5)\ (1,\ 6)\ (1,\ 7)\ (1,\ 8) \\ (2,\ 5)\ (2,\ 6)\ (2,\ 7)\ (2,\ 8) \\ (3,\ 5)\ (3,\ 6)\ (3,\ 7)\ (3,\ 8) \\ (4,\ 5)\ (4,\ 6)\ (4,\ 7)\ (4,\ 8) \end{bmatrix}$$

where each ordered pair is contained in parentheses and the entire Cartesian product is enclosed within the brackets. For ease in reading, we have grouped the Cartesian product set of ordered pairs of $A \times B$ so that the left member of the pair, or first element, is the same for each horizontal row and the right member of the pair, or second element, is the same for each vertical column.

Definition 8-1 stated that both x and y may come from either different sets or the *same set.* If

$$A = \{a, b\}$$

then

$$A \times A = \{(a, a), (a, b), (b, a), (b, b)\}$$

Exercises 8-1

1. If $A = \{1, 2\}$ and $B = \{4, 5\}$, write the Cartesian products for
   a. $A \times B$
   b. $B \times A$
   c. $A \times A$

2. Given $A = \{1, 2, 3\}$, $B = \{1, 2\}$, $C = \{2, 3\}$. Let $U = \{1, 2, 3\}$ and tabulate
   a. $A \times (B \cup C)$
   b. $B \times (A \cap C)$
   c. $(A \cap B) \times (A \cap C)$
   d. $(A \times B) \cup (A \times C)$
   e. $(A \times B) \cap (A \times C)$
   f. $(\overline{B} \cup A) \times C$

3. a. In Exercise 2 are b) and e) the same?
   b. In Exercise 2 are b) and c) the same?

   c. If set B has n elements and set C has m elements, how many elements are in $B \times B$? $C \times B$? $C \times C$?
   d. Which is true:

$$A \times (B \cap C) = (A \times B) \cap (A \times C)$$

   or

$$A \times (B \cap C) = (A \cup B) \times (A \cup C)$$

   where A, B, and C are not empty sets?
4. Given $A = \{2, 4, 5\}$. Draw a lattice picture similar to Fig. 8-2 of $A \times A$.

_____ 8-2 FIRST DEGREE EQUATIONS

Definition 8-3

A first degree equation in two variables is an equation that can be written in the form

$$ax + by + c = 0$$

where a and b cannot both be zero at the same time. This kind of equation can be graphed as a straight line. Because a straight line can be determined by two unique points, only two ordered pairs are needed to graph the straight line. Thus, the equation for the straight line in Fig. 8-3 is

$$2x + 3y = 6$$

If $y = 0$, then $x = 3$, and if $x = 0$, then $y = 2$.

   If the first and second pairs of coordinates have zero for x and y respectively, the solution set is easy to find. For example, in (0, 2) and (3, 0) we can quickly plot the line shown in Fig. 8-3 by setting each variable, in turn, to zero and determining the value of the other variable.

   The x-coordinate of the point at which the line crosses the x-axis is called the x-intercept, and the y-coordinate of the point at which the line

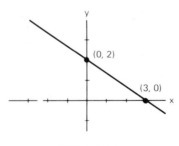

FIGURE 8-3

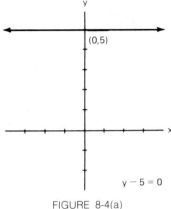

FIGURE 8-4(a)

crosses the y-axis is called the y-intercept. In Fig. 8-3, the x-intercept is 3 and the y-intercept is 2.

EXAMPLE 1   The equation $y - 5 = 0$ can be expressed in the general format $ax + by + c = 0$ by writing $0x + 1y + (-5) = 0$.

Note that for every x, $y = 5$. The graph of this equation, a line parallel to the x-axis and 5 units above it, is shown in Fig. 8-4a.

The ordered pairs for this equation are found as $(x, 5)$, where $x \in R$. Only one value is found for y, but any value can be assigned to x.

EXAMPLE 2   The equation $x - 3 = 0$ can be expressed in the general format of $ax + by + c = 0$ by writing $1x + 0y - 3 = 0$.

Note that for every y, $x = 3$. The graph of this equation, a line parallel to the y-axis and 3 units to the right of it, is shown in Fig. 8-4b.

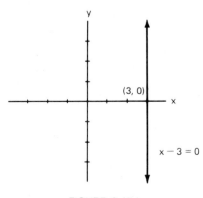

FIGURE 8-4(b)

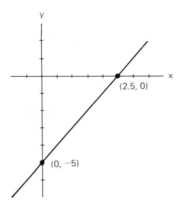

FIGURE 8-5

The ordered pairs for this equation are found as (3, y) where $y \in R$. Only one value is found for x, but any value can be assigned to y.

EXAMPLE 3 The equation $y = 2x - 5$ can be expressed in general format as $-2x + 1y + 5 = 0$. When $x = 0$, $y = -5$, and we have the ordered pair (0, -5). When $y = 0$, $x = 2.5$. These two points, (0, -5) and (2.5, 0), are enough to determine the graph of the line $y = 2x - 5$, Fig. 8-5.

8-3 RELATIONS AND FUNCTIONS

Definition 8-4

A relation is any subset of $U \times U$, where U is a non-empty set.

We can show examples of a relation in each of the following, where $U = \{-5, -2, -\frac{1}{2}, 3, 4, 6\}$:

1. In the listing format (where R means relation):
$$R = \{(3, -\tfrac{1}{2}), (4, -2), (6, -5)\}$$

2. In set builder notation:
$$R = \{(x, y)\,|\,3x + 2y = 8, \text{ where } x = 3, 4, \text{ and } 6\}.$$

3. In tabular form:

x	y
3	$-\frac{1}{2}$
4	-2
6	-5

4. In graphic form (Fig. 8-6):

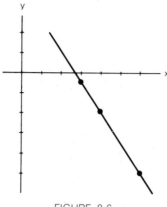

FIGURE 8-6

Because the members of the ordered pairs in the relation R are elements in U, the relation is a subset of $U \times U$.

Definition 8-5

The set of all the first coordinates of the ordered pairs in a relation is called the *domain* of the relation. The set of all the second coordinates of the ordered pairs is called the *range* of the relation.

EXAMPLE 1 $R = \{(2, -3), (3, -4), (4, -5)\}$, where $U = \{-5, -4, -3, -2, -1, 0, 1, 2, 3, 4, 5\}$.

The set of all first members of ordered pairs in this relation is $\{2, 3, 4\}$; therefore, this set is denoted as the domain.

The set of all second members of ordered pairs in this relation is $\{-3, -4, -5\}$; therefore, this set is denoted as the range.

EXAMPLE 2 $U = \{0, 1, 2\}$. Draw the graph of the relation $R = \{(x, y)|x < y\}$.

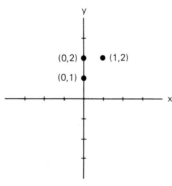

FIGURE 8-7

Since $U \times U = \{(0, 0), (0, 1), (0, 2), (1, 0), (1, 1), (1, 2), (2, 0), (2, 1),$ $(2, 2)\}$, then $R = \{(0, 1), (0, 2), (1, 2)\}$, because these are the only coordinates that satisfy $x < y$. (See Fig. 8-7.)

Definition 8-6

An *inverse* of a relation is another relation obtained by the interchange of the order of the coordinates of all the ordered pairs of the original relation. The inverse of a relation R is denoted by R^{-1}.

EXAMPLE   If $R = \{(4, 5), (-2, 3), (0, 1)\}$, then $R^{-1} = \{(5, 4), (3, -2), (1, 0)\}$.

Before defining a function, consider a car moving at 55 miles per hour. To know the distance traveled, we must first know the total amount of time it was driven. The idea of one variable, such as distance, depending upon another variable, such as time, is such a common occurence in mathematics that there is a special name for it, a *function*. We can state that when two variables are so related that assigning a value to one determines a value for the other, the second variable is said to be a function of the first. More formally, we have.

Definition 8-7

A *function* is a relation in U such that for each first coordinate, there is one and only one second coordinate. Thus, a function is a special type of relation that associates each element, x, in its domain with one and only one element, y, in its range.

EXAMPLE 1   Let

$$U = \{1, 2, 3, 4\}$$

Then

$$S = \{(1, 2), (2, 3), (3, 4)\} \text{ is a function.}$$

Table 8-1

x	y
1	2
2	3
3	4
3	5

EXAMPLE 2   Table 8-1 does not describe a function because 3 appears twice as a first coordinate. (Compare to Definition 8-7.)

EXAMPLE 3   Fig. 8-8 is a graph of a function. *Any* nonvertical straight line is a graph of a function. (Figs. 8-4(b) and 8-7 are not graphs of functions—they are graphs of relations.)

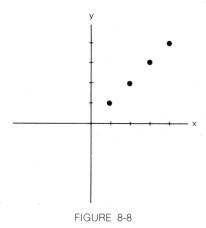

FIGURE 8-8

EXAMPLE 4   Let

$$U = \{x, y \mid x \in S, y \in S\}$$

Then

$$S = \{(x, y) \mid y = 2x + 5\}$$

is a function since any x gives a unique value for y.
    To simplify all of this, we will write

$$f(x) = 2x + 5$$

where the symbol $f(x)$ is read as "the function of x" or "f of x."

8-3.1   Notation

    The symbol $f(x)$ can be used to specify the y-coordinate of the or-
dered pair whose first member is x. Therefore, $y = f(x)$. We can now write
$(x, f(x))$ is in f, or the function. In $f = \{(x, y) \mid y = 5 - 2x\}$, if $x = 3$, then
$y = f(3) = -1$. The value for y is obtained by substituting 3 for x in
$y = 5 - 2x$. Hence, $(3, -1)$ is a member of the function f.

Definition 8-8

    The *inverse function* f^{-1} is the set of ordered pairs obtained by
    the interchange of the coordinates of all of the ordered pairs
    of the original function, such that the result itself is a function.

The result of interchanging coordinates is a function only if Definition 8-7
holds true for that result. Thus, $\{(-1, 1), (0, 0), (1, 1)\}$ is a function, but
$\{(1, -1), (0, 0), (1, 1)\}$ is *not* a function.

EXAMPLE 1   In $f_1 = \{(3, 6),(4, 7), (5, 8)\}$, the interchange of coordinates
yields $\{(6, 3), (7, 4), (8, 5)\}$.

Because Definition 8-7 holds true after the interchange of coordinates, f_1 has an inverse function.

Fig. 8-9 contains the graph of f_1 and f_1^{-1}.

EXAMPLE 2   In $f_2 = \{(0, 4), (1, 4), (2, 5)\}$, the interchange of coordinates yield $\{(4, 0), (4, 1), (5, 2)\}$. As an exercise plot the points in f_2 and f_2^{-1}.

Because Definition 8-7 does not hold true after the interchange of coordinates, f_2 does not have an inverse function.

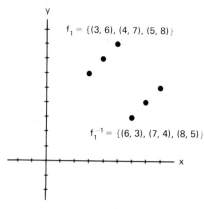

FIGURE 8-9

Exercises 8-2

1. Use the listing format and denote the domain and range for
   a. $\{(3, 2), (-1, 3)\}$
   b. $\{(-2, 1), (-1, 0), (0, -1), (1, -2)\}$
   c. $\{(x, y) \mid x + 3y = 1\}$, where the Universal Set $= \{-2, -3, -4, 1, 2, 3\}$
   d. What are the members of the relation in part c?

2. a. Draw the points for $U \times U$, where $U = \{-2, -1, 0, 1, 2\}$.
   b. Graph $x - y = 1$, where $U = \{-2, -1, 0, 1, 2\}$.

3. Find the inverse for
   a. $\{(3, -2), (0, 1), (1, 0)\}$
   b. $\{(1, 0), (0, -1), (1, 1), (-1, 0)\}$

4. Identify which of the following define a function.
   a. $x + y = 1$
   b. $y = 3x + c$
   c. $y = kx$
   d. $x = 4$

   e.

x	y
-1	2
0	3
1	4
1	5

5. Given the domain $D = \{1, 3, 5\}$ and $f(x) = 3 - 2x$, find the range.
6. Given $f(x) = x + 2$, find the domain when
   a. $f(x) = 0$
   b. $f(x) = 2$
   c. $f(x) = -1$
7. Given $R = \{(1, 2), (2, 3), (3, 4)\}$, find R^{-1}.
8. Does $R = \{(x, y)|y = x + 3\}$ have an inverse function?

_____ **8-4 SLOPE AND y-INTERCEPT**

There are two fundamental properties of a line on the Cartesian coordinate system, its *slope* and *y-intercept*. The slope of a line is measured as the ratio of the difference between the y coordinates of any two points to the difference between the x-coordinates of the same two points. Letting m represent the slope, we can write

$$m = \frac{y_1 - y_2}{x_1 - x_2} = \frac{\text{difference in } y\text{-coordinates}}{\text{difference in } x\text{-coordinates}}$$

where (x_1, y_1) and (x_2, y_2) are two points on the line. We can also write

$$\text{slope} = \Delta y / \Delta x$$

where the Greek letter Δ means the "change in" or "difference in."
If the line is parallel to the x-axis, $y_1 - y_2 = 0$. (Why?) Then

$$m = \frac{0}{x_1 - x_2} = 0.$$

If the line is parallel to the y-axis, $x_1 - x_2 = 0$. (Why?) Then

$$m = \frac{y_1 - y_2}{0}$$

is undefined, since division by zero is undefined.

EXAMPLE 1 Find the slope of a line that joins the points (5, 4) and (7, 2). (See Fig. 8-10.)

$$\text{slope} = \frac{\Delta y}{\Delta x} = \frac{4 - 2}{5 - 7} = -1$$

EXAMPLE 2 Find the slope of a line that passes through the origin and $(-4, 3)$. (Remember that the origin is denoted as (0, 0).)

$$m = \frac{3 - 0}{-4 - 0} = \frac{3}{-4}$$

Using

$$m = \frac{y_1 - y_2}{x_1 - x_2}$$

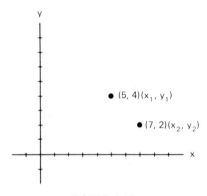

FIGURE 8-10

we can write this expression in another manner, thus:

$$y_1 - y_2 = m(x_1 - x_2)$$

Consider the graph of a line with the slope passing through a given point on the y-axis and having the coordinates $(0, b)$. Substituting 0 and b into the formula, we obtain

$$y_1 - b = m(x_1 - 0) \text{ or } y_1 = mx_1 + b.$$

Deleting the subscripts, we can write

$$y = mx + b$$

The y-intercept is the coordinate where the graph crosses the y-axis and is denoted here by the letter b. The line has a slope m, the coefficient of x. This equation is called the *slope-intercept* form. For example, $3x + 4y - 8 = 0$ can now be written as

$$y = -\tfrac{3}{4}x + 2$$

The slope of this line is $-\tfrac{3}{4}$, and the y-intercept is 2.

Exercises 8-3

1. Draw the graph of $x + y = 5$. To help plot the graph, obtain several values that satisfy the equation. Some of these are:

x	y
5	0
4	1
3	2

2. Draw the graph of $3x - 2y = 12$.
3. Draw the graph of $-4x + y = 8$.
4. Draw the graph of $2x - 4 = 0$.
5. Draw the graph of $2x - 3y = 0$.

Write each of the equations in Exercises 6 through 14 in the form $y = mx + b$.

6. $x + 2y = 1$
7. $4x - 3y = 7$
8. $-5x = y - 5$
9. $x - y = 3$
10. $5x + y = 25$
11. $3x - 4y = -12$
12. $x - y = 0$
13. $2x - y = 7$
14. $x = -y + 1$
15. Draw the graph for a line with the same slope as $2x - y = 3$, but passing through the origin.
16. Draw a line through (0, 3) with the slope found in $x - y = 3$.

Use the values given below to write an equation of each line in the format $y = mx + b$.

17. $(2, -3)$, $m = \frac{1}{2}$
18. $(0, 0)$, $m = 2$
19. $(2, 2)$, parallel to the y-axis
20. $(-2, -2)$, parallel to the x-axis
21. $(4, -5)$, $m = -\frac{3}{4}$

Write a linear equation in the format $ax + by + c = 0$ for each of the following pairs of coordinates.

22. (1, 3) and (3, 1)
23. (1, 2) and (-2, 3)
24. (4, 0) and (0, 3)
25. (-4, 3) and (-4, -3)

———————— 8-5 GRAPHING LINEAR INEQUALITIES

Consider the graph of the inequality $x > y$. If we wished to select some ordered pairs to satisfy this inequality, we could write:

x	y
3	2
-2	-3
1	0
-4	-6
0	-3

Drawing a dotted line for $x = y$ makes it easy to locate these ordered pairs. Note that all points for $x > y$ lie in the shaded area in Fig. 8-11.

Consider the following points:

a. $(-2, 1)$ where $1 > -2$. These coordinates lie outside the shaded area.
b. $(2, 0)$ where $2 > 0$. These coordinates lie in the shaded area.

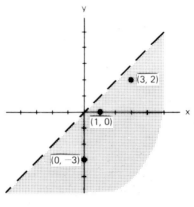

FIGURE 8-11

The diagonal line in Fig. 8-11 has been broken to indicate that $x = y$ contains points which are not part of the solution set.

Using the general form for a straight line, $y = mx + b$, we can graph the straight line for $y = -3x + 6$, where $m =$ the slope $= -3$, and $b =$ the y-intercept $= 6$.

If we set $y = 0$, then $0 = -3x + 6$, or $x = 2$; hence, the x-intercept $= 2$. Connecting the points at $y = 6$ and $x = 2$, we have line *AB* in Fig. 8-12.

Now consider $y \geq -3x + 6$.

Because the symbol $\geq$ means greater than *or equal to,* the line in Fig. 8-12 is not broken but solid, and all the points on this line, as well as those points found in the shaded area, belong to the graph for $3x + y \geq 6$. (Note that the origin, or point $(0, 0)$, is outside the shaded area.)

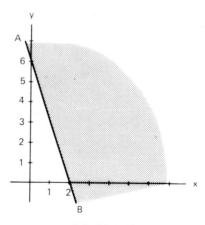

FIGURE 8-12

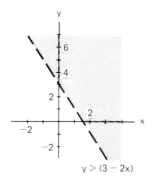

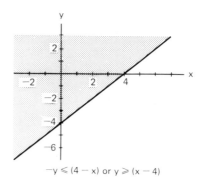

FIGURE 8-13

The solution set that contains ordered pairs common to *two* inequalities can be located in a region of the plane where two shaded areas overlap.

For example, in:

$$y > (3 - 2x)$$
$$-y \leq 4 - x$$

we can plot and shade as shown in Fig. 8-13. To remove the − sign before y, we multiply the second inequality by − 1

Thus $-1(-y \leq 4 - x)$
yields $y \geq x - 4$

and we can now write both inequalities as

$$y > 3 - 2x$$
$$y \geq x - 4$$

Multiplying an inequality by − 1 causes the inequality sign to be reversed.

Combining these two graphs, we get Fig. 8-14.

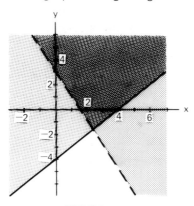

FIGURE 8-14

The part of the graph in which the two shaded areas overlap contains the solution set for the inequalities $y > (3 - 2x)$ and $y \geq (x - 4)$.

Given:

$$\{(x, y)|x > 3\}$$
$$\{(x, y)|x + y \geq 4\}$$

Is there a solution set? To find out, we can combine graphs of these sets of ordered pairs in Fig. 8-15.

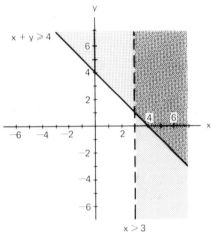

FIGURE 8-15

The overlapping of the shaded areas contains the solution set for

$$x > 3$$
$$x + y \geq 4$$

As an exercise, determine if there is a solution set for

$$\{(x, y)|\ x \leq 5\}$$
$$\{(x, y)|y > -2\}$$

Exercises 8-4

Draw the graph and shade the appropriate area for Exercises 1 through 5.

1. $x - y > 2$
2. $x + y < 4$
3. $3x + 2y \geq 6$
4. $y \geq 4x - 3$
5. $3x < 8 - y$

Graph and indicate the area that represents the solution set for Exercises 6 through 9.

6. $x + y > 4$
   $x - y < 5$
7. $3x + 2y > 7$
   $y + 2x < 8$
8. $y - x > 5$
   $4x + y \geq 7$
9. $2x + y \geq 6$
   $x > 2$

Draw the graphs for the following.

10. $x = |y|$
11. $y = |x + 2|$
12. $|y| > 1$

Hint: in $y = |x|$ we have for positive x, $y = x$; for negative x, $y = -x$; and the graph for $y = |x|$ is

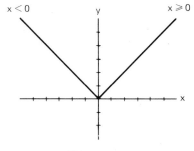

FIGURE 8-16

13. Suppose $y \geq -3x + 6$, and
    1. y must always be equal to or greater than zero.
    2. x is the only input variable, and x must never be greater than 3.
    3. values for x are to be input in ascending order, beginning with the negative values.
    4. the only outputs required are the coordinates x and y.

We can illustrate the appropriate algorithm for this situation using pseudocode, or brief statements in English that closely resemble the Basic language code and make the algorithm easier to describe. (Note: the "DO the following WHILE" statement in pseudocode means that we are to process all values for x WHILE x is less than 3. When x becomes greater than 3, the program halts.)

Using the pseudocode as a guide, complete the BASIC program statements to output x and y when $y \geq 0$ and until $x > 3$. Remember that any time $y < 0$, we will not print the coordinates.

PSEUDOCODE
Set $m = -3$
Set $b = 6$
Input x
DO the following WHILE x is not greater than 3
    Set $y = mx + b$
    IF y is equal to or greater than zero
        Then output x, y
    End of IF statement
    Input x
End of DO WHILE
Stop

BASIC
005 LET $M = -3$
010 LET $B = 6$
015 INPUT X
020 IF $X > 3$ THEN GO TO 040
025 _____
030 _____
032 _____
035   INPUT X
037 GO TO 020
040 STOP
045 END

8-6 SIMULTANEOUS SOLUTION OF LINEAR EQUATIONS

When a group of algebraic equations can be satisfied by the same values of the variables, these equations are called *simultaneous equations*, i.e., they are simultaneously true.

Definition 8-9

The *solution set* for two equations in two unknowns x and y is the set of ordered pairs (x, y) that satisfy both equations.

For example, $x = 2y + 3$ and $x + 3y = 8$ are two equations with $x = 5$ and $y = 1$. The solution set is $\{(5, 1)\}$. Solving these equations involves

1. Eliminating one of the variables.
2. Finding the value for the remaining variable.
3. Substituting this value into one of the original equations to find the value of the other variable.

4. Substituting the values for the variables into both original equations to check or verify the results.

We will solve simultaneous equations by illustrating the procedure above and using either 1) substitution or 2) addition and subtraction.

8-6.1 Substitution

Simultaneous equations can be solved when one of the variables in one of the two equations can be replaced by its value in the other equation, as, for example, in

$$x = 2y + 3$$
$$x + 3y = 8$$

Since $x = 2y + 3$ in the first equation, we can replace x by $2y + 3$ in the second equation to obtain

$$(2y + 3) + 3y = 8$$
$$5y = 5$$
$$y = 1$$

Substituting 1 for y in the first equation yields

$$x = 2 + 3$$
$$x = 5$$

Results can be verified by using these values for x and y.

$$5 = 2 + 3$$
$$5 + 3 = 8$$

8-6.2 Addition, Subtraction

Another way of solving simultaneous linear equations is by addition or subtraction. For example, we can solve the two equations $2x + y = 7$ and $x - y = -4$ by addition using Theorem 0-1,*

$$
\begin{array}{rl}
2x + y &= 7 \\
+\ \ x - y &= -4 \\
\hline
3x\ \ \ \ &= 3 \\
x\ \ \ \ &= 1
\end{array}
$$

---

* Actually, a generalization of Theorem 0-1, where if $a = b$ and $c = d$, then $a + c = b + d$.

To obtain a value for y, substitute $x = 1$ in the first equation:

$$2(1) + y = 7$$
$$y = 5$$

With these values of x and y, we can verify that they are solutions:

$$2 + 5 = 7$$
$$1 - 5 = -4$$

Consider now the equations

$$2x + 3y = 11$$
$$x + 4y = 8$$

Since none of the coefficients of x or y is the same, we will multiply the terms in the second equation by an appropriate coefficient and add or subtract to eliminate one of the variables. Refer to Theorem 0-3 (page 12). Multiplying the second equation by 2 and recopying the first equation, we obtain

$$
\begin{array}{r}
2(x) + 2(4y) = 2(8) \\
- \quad 2x + 3y = 11 \\
\hline
5y = 5 \\
y = 1
\end{array}
$$

Subtracting the first equation from the second, we obtain $5y = 5$, or $y = 1$. Continuing as before, we obtain $x = 4$.

8-6.3 Graphing

Graphing is a third method used to solve simultaneous equations. The following steps assist in solving simultaneous equations by graphing:

1. Draw the lines for each equation by finding the slopes and the y-intercepts.
2. Identify the coordinates of the point at which the two lines intersect. (Note: Equations whose graphs intersect at one point are called *consistent*, or *independent*, equations.)
3. Verify the coordinates by substituting their values into the original equations.

EXAMPLE Solve graphically

$$y = 4 - \tfrac{1}{2}x$$
$$y = 1 + x$$

These equations are graphed in Fig. 8-17.

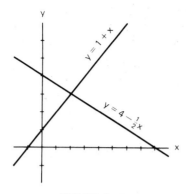

FIGURE 8-17

Identify the coordinates at the point where the two lines intersect as $x = 2$, $y = 3$. Verify as before:

$$2 + 2(3) = 8$$
$$2 - 3 \quad = -1$$

Exercises 8-5

Solve Exercises 1 through 9 without the use of graphs.

1. $4x + 2y = 8$
   $x - y = 5$
2. $x = y + 2$
   $3y + 5x = 8$
3. $x + 2y = 6$
   $3x + y = 3$
4. $11x + 33y = 55$
   $2x + 3y = 8$
5. $x + y = 10$
   $2x - y = 10$
6. $5x + y = 14$
   $2x = y + 2$
7. $\frac{1}{3}x + \frac{1}{2}y = 2$
   $\frac{1}{6}x + \frac{1}{3}y = \frac{2}{3}$
8. $x = y$
   $2x + y = 5$
9. $y = x + 3$
   $-x = y - 2$
10. The sum of the digits of a two-digit number is 8. If the digits are reversed, the new number is greater than the original number by 18. Find both digits.

11. A pleasure boat makes a ten-mile downstream trip in 1.5 hours. The upstream trip takes 3 hours. Find the rate of the current and the rate of the boat in still water.

Solve problems 12 through 16 by the use of graphs.

12. $x = y$
       $x + y = 8$
13. $2x + 2y = 0$
    $3x + 2y = 3$
14. $5x - 3y = 6$
    $x - 2y = 1$
15. $y = 3x + 2$
    $x = 2y + 1$
16. $3x + 6y = 33$
    $x = 3$

17. A note is discounted at 4%. Another note is discounted at 6%. The annual income from both notes is $420. If the amounts invested at each rate had been interchanged, the annual income would have been $480. What were the original amounts?

18. The cost to produce two fuel injection systems and three pollution-free electrical engines is $2,500, while the cost of producing three fuel injection systems and two pollution-free electrical engines is $3,000. Find the cost of one fuel injection system and one pollution-free electrical engine.

   a. Solve by simultaneous equations
   b. Solve by graphing

8-7 SPECIAL CASES IN FIRST DEGREE EQUATIONS

In the preceding examples we have assumed that a solution exists and have located this solution as the intersection of two lines on a graph. However, there are two cases where either many solutions exist or no solutions exist; we will examine them here.

Because the graph of every linear equation in two unknowns is a straight line, we are able to describe these two special situations graphically.

8-7.1 Dependent Systems of Equations

Consider

$$2x - y = 4$$
$$-6x + 3y = -12$$

Trying to solve this system of equations by multiplication of the first equation by 3 and then adding, we obtain

$$\begin{array}{r} 6x - 3y = 12 \\ + \ -6x + 3y = -12 \\ \hline 0 = 0 \end{array}$$

The statement $0 = 0$ is always true for all x and y which satisfy both of these equations. Therefore, all solutions for one equation in this system are solutions for the other. These equations are graphed in Fig. 8-18.

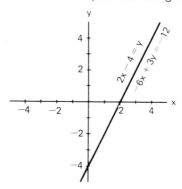

FIGURE 8-18

We note that the line for $2x - y = 4$ and the line for $-6x + 3y = -12$ are actually the same line. Further, we note that the slope is 2 and y-intercept is -4 for both lines, so that both lines can be described as coincident. This means that for $x \in R$ and $y \in R$, any pair (x, y) that is a solution for one equation is a solution for the other equation. Thus, every point on the coincident lines is a solution for the equations. Systems of equations that have this property are called *dependent*.

If we multiply the first equation, $2x - y = 4$, by the constant -3, we get the second equation, $-6x + 3y = -12$, and in general we see that where k equals some constant we can write a system of dependent equations as

$$ax + by + c = 0$$
$$kax + kby + kc = 0$$

If we solve for y in the first equation, we get

$$y = \frac{-(c + ax)}{b}$$

Substituting this expression in the second equation, we have

$$kax + kb \frac{-(c + ax)}{b} + kc = 0$$
$$kax - kc - kax + kc = 0$$
$$0 = 0$$

Thus, we find that the method of substitution will not work for dependent systems: in dependent systems, one equation is a multiple of the other, and the number of possible solutions is infinite.

8-7.2 Inconsistent Systems of Equations

In:

$$x + 3y = 5$$
$$2x + 6y = 8$$

Trying to solve this system of equations by multiplication and subtraction, we get

$$2x + 6y = 10$$
$$\underline{2x + 6y = 8}$$
$$0 = 2$$

The statement $0 = 2$ is a contradiction. The solution set is then the *null* set. (See page 65.)

Now suppose instead that we try to solve these equations by the method of substitution. From the first equation, we get

$$x = 5 - 3y$$

Substituting this value for x in the second equation, we obtain

$$2(5 - 3y) + 6y = 8$$
$$10 - 6y + 6y = 8$$
$$10 = 8$$

which is another contradiction.

Again, our methods for solving simultaneously will not work. The equations are graphed in Fig. 8-19.

Note that the line for $x + 3y = 5$ and the line for $2x + 6y = 8$ are parallel. We define two lines as parallel if and only if they have the same slope. Further, note that although the y-intercepts are $\frac{5}{3}$ and $\frac{4}{3}$, respectively, the slopes are the same, or $-\frac{1}{3}$. Because parallel lines never intersect, there can be no common solutions (x, y) for this system of equations where $x, y \in R$. Systems of equations that have this property are called *inconsistent equations*.

Note that in $x + 3y$ and $2x + 6y$ the coefficients of x and y are multiples of each other. Therefore we can write

$$ax + by + c = 0$$
$$kax + kby + d = 0$$

We cannot write $kax + kby + kc$ in the second equation because $kc \neq d$.

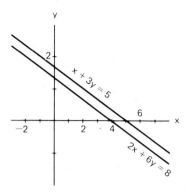

FIGURE 8-19

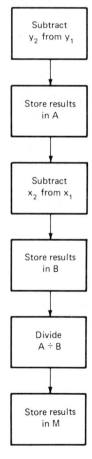

FIGURE 8-20

Exercises 8-6

Identify the following equations. Are they dependent, inconsistent, or independent? (Determine if they are coincident, parallel, or if they intersect at one point.)

1. $\begin{cases} 2x - y = 3 \\ -4x + 2y = -6 \end{cases}$

2. $\begin{cases} y = 11 + x \\ x + 2y = 8 \end{cases}$

3. $\begin{cases} x + y = 1 \\ 2x + 2y = 3 \end{cases}$

4. $\begin{cases} 5x + y = 2 \\ -15x = 3y - 6 \end{cases}$

5. a. Describe the activities in the flowchart shown in Fig. 8-20.
   b. Write the associated Fortran statements. Modify the flowchart to test for
      1. Dependent equations.
      2. Inconsistent equations.

Code your results in Fortran. (Refer to the IF statement in Chapter 6 if necessary.)

6. Write a definition for

   a. Consistent equations.
   b. Inconsistent equations.
   c. Dependent equations.

Write the definitions in terms of the graphs of these types of equations.

9

QUADRATIC AND LOGARITHMIC FUNCTIONS

In Chapter 8 we were concerned with linear functions and first degree equations. Data processing problems such as

—the relationships between costs, profits and the production of goods
—compound interest programs using logarithms
—statistical measurements

require equations that are usually expressed as nonlinear functions, whose graphs are not straight lines.

Before considering commercial applications of nonlinear functions, we will discuss a nonlinear function called the *quadratic function*.

Definition 9-1

The function f for which $f(x) = ax^2 + bx + c$ is a *quadratic function* for $x \in R$, where a, b, c are constants in the set of real numbers, $a \neq 0$.

(Note: The value of x when $f(x) = 0$ is called the *zero* of the function.)

9-1 SOLVING QUADRATIC EQUATIONS

9-1.1 Factoring

To find the solution set for $x^2 - 3x + 2 = 0$, factor the left side of the equation to obtain $(x - 1)(x - 2) = 0$.

When the product of two factors is zero, at least one of the factors must be zero. (If $a \cdot b = 0$, then $a = 0$, or $b = 0$ or both; see page 12.)

Therefore, either $(x - 1) = 0$ or $(x - 2) = 0$, or both $x - 1 = 0$ and $x - 2 = 0$.

Then $x = 1$ and/or $x = 2$. Since both $x = 1$ and $x = 2$ satisfy the equation, the solution set for $x^2 - 3x + 2 = 0$ is $\{1, 2\}$.

9-1.2 Completing the Square

In some cases where the coefficient of x^2 is 1 (*i.e.*, $a = 1$) we may find the solution set of a quadratic equation by *completing the square*.

To get $x^2 + mx = c$ in the form where the left hand side is a perfect square, add $m^2/4$ to both sides of the equation. Then

$$x^2 + mx + \frac{m^2}{4} = c + \frac{m^2}{4}$$

$$\left[x + \frac{m}{2}\right]^2 = c + \frac{m^2}{4}$$

where the left-hand side is a perfect square. One can determine $m^2/4$ by taking half the coefficient of x and squaring. This process is called "completing the square."

In $x^2 + 6x + 9$, we can write $x^2 + 6x + 9 = (x + 3)^2$, and $(x + 3)^2$ is the square for $x^2 + 6x + 9$. Therefore, $x^2 + 6x + 9$ is a perfect square trinomial.

As a sample problem, assume that only $x^2 + 6x$ were given and the question was "What value for c must be added to $x^2 + 6x$ to find a perfect square trinomial?" Another way this problem may be stated is "Find c in $x^2 + 6x + c$, where $x^2 + 6x + c$ is a perfect square trinomial."

By dividing the coefficient of x by 2 and squaring the result, we obtain

$$\left(\frac{6}{2}\right)^2 = 9 = c$$

Thus, we obtain the quadratic equation

$$x^2 + 6x + 9 = 0$$

Since $x^2 + 6x + 9 = (x + 3)^2$, $x = -3$ or $\{-3\}$ is the solution set for this equation.

As another example, to find the zeros of the function $f(x) = x^2 + 6x + 3$, set $f(x) = 0$. Then add 6 to both sides in completing the square:

$$x^2 + 6x + 9 = 6$$
$$(x + 3)^2 = 6$$
$$x + 3 = \pm\sqrt{6}$$
$$x = \pm\sqrt{6} - 3$$

9-1.3 The Quadratic Formula

The quadratic formula is another method used to solve for x. To solve $ax^2 + bx + c = 0$ for x, we can express the solution set of a quadratic formula in terms of its coefficients ($a \neq 0$). The solution set for x is

$$\left\{ \frac{-b + \sqrt{b^2 - 4ac}}{2a}, \frac{-b - \sqrt{b^2 - 4ac}}{2a} \right\}$$

expressed in terms of the coefficients in the quadratic equation. By definition 9-1, the coefficients are constants in the set of real numbers. Therefore, they may be substituted for in the quadratic formula. (Proof of the quadratic equation is given in Appendix B.)

EXAMPLE   Using the quadratic formula, find the solution set for

$$2x^2 + 3x + 1 = 0$$
$$x = \frac{-3 \pm \sqrt{3^2 - 4 \cdot 2 \cdot 1}}{2 \cdot 2}$$
$$x = \frac{-3 \pm 1}{4}$$

In this case the solution set is $\{-\frac{1}{2}, -1\}$.

Note that in the quadratic formula the radicand, $b^2 - 4ac$, is called the *discriminant*. (See page 316.) Because the discriminant is a radicand, it can be used to determine whether the roots of the quadratic equation will be real or complex, equal or unequal. The three possible cases are:

1. If $b^2 - 4ac > 0$, the roots are real and unequal.

EXAMPLE   Let

$$a = 1, b = 2, c = -1$$

Then

$$2^2 - 4 \cdot 1 \cdot (-1) = 8, \text{ and } 8 > 0$$

2. If $b^2 - 4ac = 0$, the roots are real and equal.

EXAMPLE   Let

$$a = 1, b = 2, c = 1$$

Then

$$2^2 - 4 \cdot 1 \cdot 1 = 0$$

3. If $b^2 - 4ac < 0$, the roots are complex (see page 3).

EXAMPLE   Let

$$a = 1, b = 2, c = 3$$

Then

$$2^2 - 4 \cdot 1 \cdot 3 = -8 \text{ and } -8 < 0$$

Exercises 9-1

Solve for x in

1. $3 - [(x + 2)(x - 2)] = 6$
2. $3x^2 + 9x = 27$

Write the quadratic equations in standard form for Exercises 3 and 4.

3. $\dfrac{-5 + \sqrt{5^2 - 4 \cdot 3 \cdot 4}}{2 \cdot 3}$

4. $\dfrac{-\frac{1}{2} + \sqrt{(\frac{1}{2})^2 - 4 \cdot 1 \cdot 2}}{2}$

5. Solve using the quadratic formula: $2x^2 + 5x - 1 = 0$
6. Given $x^2 + 10x$, complete the square and find a value for c that yields a perfect square trinomial.
7. The quadratic formula yields two roots for x:

$$\frac{-b + \sqrt{b^2 - 4ac}}{2a} \quad \text{and} \quad \frac{-b - \sqrt{b^2 - 4ac}}{2a}.$$

a. Find the sum of these two roots.
b. Find the product of these two roots.

8. Find the sums and products of the roots in

a. $(x + 3)(x - 2)$
b. $x^2 + x - 42$

9. Write one Fortran statement to test for both a negative discriminant and a discriminant of zero.
10. Taking the square root of real numbers is performed in Fortran by a subroutine or small program provided by the manufacturer. To take the square root of real numbers, the function name SQRT must be entered, followed by the real number enclosed in parentheses. If we wanted the square root of 625, for example, we would simply write

$$X = \text{SQRT (625.)}$$

where the variable X is replaced by 25. To take the square root of a variable named Y we would write

$$A = \text{SQRT (Y)}$$

Assume that we have used the coding in the answer to problem 9 to make certain that we will only take the square root of real numbers greater than zero. Write the Fortran statement that will solve for x in $ax^2 + bx + c = f(x)$, where $f(x) = 0$.

9-2 ORDERED PAIRS AND GRAPHING

With linear equations in two variables, we found solutions to be ordered pairs (x, y). Quadratic equations in two variables also have solutions that are ordered pairs. A general format for this type of quadratic equation is $y = ax^2 + bx + c$.

In $y = x^2 - 3$ (where $a = 1$, $b = 0$ and $c = -3$), we can create a partial table of *replacement values* for x and y, or $f(x) = ax^2 + bx + c$. Ordered pairs are found by arbitrarily assigning values to x and computing the appropriate values for y.

Table 9-1

x	$f(x) = y$
-3	6
-2	1
-1	-2
0	-3
1	-2
2	1
3	6

Assigning -3 to x, we find

$$y = (-3)^2 - 3$$
$$y = 6$$

and the ordered pair for this value of x is $(-3, 6)$. In a similar manner, we can find other values for y in Table 9-1 and write the ordered pairs as follows:

$(-3, 6), (-2, 1), (-1, -2), (0, -3), (1, -2), (2, 1), (3, 6)$

Connecting these points by a smooth curve, we find a shape called a *parabola*. (See Fig. 9-1.) Note that the curve has been drawn intuitively without plotting any more than these few points on the graph.

In general, we can say that the graph of the points that make up the solution set of a parabola are in the form of $y = ax^2 + bx + c$, where a, b, c $\in R$ and $a \neq 0$. (Whenever $a = 0$, a straight line will contain the points in the solution set.) The parabola is a graph of a function, since for each x there is only one value for y.

Drawing lines parallel to the y-axis to determine the point(s) of intersection is a way of testing to see if a relation is a function or not. In Fig.

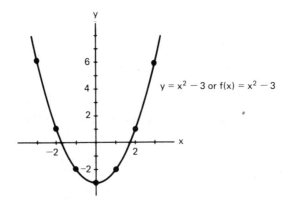

FIGURE 9-1

9-1a, the dashed line parallel to the y-axis crosses the curve twice at points A and B. Because points A and B have the same x-coordinate and different y-coordinates, this relation is not a function. More specifically, y is not a function of x in Fig. 9-1a.

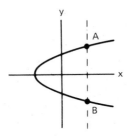

FIGURE 9-1a

In $y = x^2 - 3$ we used a value for a greater than zero. Whenever $a > 0$, the legs of the parabola extend upward. If $a < 0$, the legs of the parabola will extend downward.

$-x^2 - y - 3 = 0$

x	y
3	-12
2	-7
1	-4
0	-3
-1	-4
-2	-7
-3	-12

FIGURE 9-2

Compare the graphs of $x^2 - y - 3 = 0$ (Fig. 9-1) and $-x^2 - y - 3 = 0$ (Fig. 9-2). Note the effect of -3 in both Fig. 9-1 and Fig. 9-2. When $x = 0$ in $x^2 - y - 3 = 0$, the lowest point of the curve is at $y = -3$. When $x = 0$ in $-x^2 - y - 3 = 0$, the highest point of the curve is at $y = -3$.

9-2.1 x-Intercepts

In the graph of the function $f(x) = ax^2 + bx + c$, we can also write $y = ax^2 + bx + c$. Note that any value for x where $f(x) = 0$ will be an x-intercept (the point on the x-axis where $y = 0$) and therefore a point in the solution set of this form of the equation.

As noted earlier, the values of x for $f(x) = 0$ are called *zeros* of the function. To illustrate the zeros of the function, draw the graph for $y = x^2 - 7x + 6$ and locate $f(1) = 0$, $f(6) = 0$. (Set $f(x) = 0$; see Fig. 9-3.)

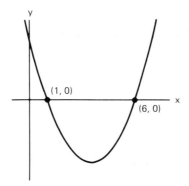

FIGURE 9-3

If one side of the quadratic equation can be factored, the zeros of the function may be located quickly. For example, in $0 = x^2 + x - 2$, we can solve as follows:

$$(x - 1)(x + 2) = 0$$
$$x = 1, \ x = -2$$

Now, using the form $y = ax^2 + bx + c$, we note that when $x = -2$ or $x = 1$, $y = 0$ (where $y = f(x)$); the zeros of the function are -2, 1, and this is precisely where the graph crosses the x-axis. Fig. 9-4 is the graph of $y = x^2 + x - 2$.

x	y
-3	4
-2	0
-1	-2
0	-2
1	0
2	4

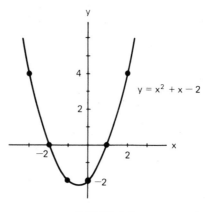

FIGURE 9-4

9-2.2 Lowest and Highest Points

A method for finding the lowest or highest point on the graph of $y = ax^2 + bx + c$ can be obtained by looking at the constants a, b, and c ($a \neq 0$).

If $a > 0$, then $(x, y) = \left(\dfrac{-b}{2a}, -\dfrac{b^2 - 4ac}{4a} \right)$ is the lowest point.

If $a < 0$, then $(x, y) = \left(\dfrac{-b}{2a}, -\dfrac{b^2 - 4ac}{4a} \right)$ is the highest point.

To see how this is derived, consider

$$y = ax^2 + bx + c, \text{ or } \frac{y}{a} = x^2 + \frac{bx}{a} + \frac{c}{a}$$

Completing the square, we get

$$\frac{y}{a} + \frac{b^2}{4a^2} = \left(x^2 + \frac{bx}{a} + \frac{b^2}{4a^2} \right) + \frac{c}{a}$$

$$\frac{y}{a} + \frac{b^2}{4a^2} - \frac{c}{a} = \left(x + \frac{b}{2a} \right)^2$$

$$\frac{y}{a} + \frac{b^2 - 4ac}{4a^2} = \left(x + \frac{b}{2a} \right)^2 \qquad (1)$$

where $\left(\dfrac{y}{a} + \dfrac{b^2 - 4ac}{4a^2} \right)$ is ≥ 0 since $\left(x + \dfrac{b}{2a} \right)^2$ is ≥ 0. (Any number squared is equal to or greater than zero.)

If $a > 0$, the parabola extends upward. In (1), then, when $x = \dfrac{-b}{2a}$,

$y = -\dfrac{b^2 - 4ac}{4a}$. Hence, if $a > 0$, then $(x, y) = \left(\dfrac{-b}{2a}, -\dfrac{b^2 - 4ac}{4a}\right)$ is the lowest point of the graph of $y = ax^2 + bx + c$. In a similar manner, if $a < 0$, then $(x, y) = \left(\dfrac{-b}{2a}, -\dfrac{b^2 - 4ac}{4a}\right)$ is the highest point of the graph of $y = ax^2 + bx + c$, and the parabola extends downward. (If $a = 0$, then the graph is for $y = mx + b$.)

Thus, $x = \dfrac{-b}{2a}$, $y = -\dfrac{b^2 - 4ac}{4a}$ are solutions for equation (1).

EXAMPLE 1   In the graph of $y = x^2 - 5x + 6$

$$x = \frac{-b}{2a} = \frac{5}{2}$$

$$y = -\frac{b^2 - 4ac}{4a} = -\frac{25 - 24}{4}$$

When $x = \frac{5}{2}$, y has a minimum value of $-\frac{1}{4}$, and the coordinates of this point are $\frac{5}{2}$, $-\frac{1}{4}$.

Exercises 9-2

Graph the following.

1. $x^2 - 7x + 6 = y$ as in Fig. 9-3; compare to the graph required in Exercise 2.
2. $-x^2 + 7x - 6 = y$
3. $x^2 - 4x + 3 = f(x)$
4. $.5x^2 + 2x = f(x)$
5. Find the x-intercepts, or zeros, of the functions in Exercises 1 and 2.
6. Graph $x^2 = y$. (Is this a function?)
7. Graph $y^2 = x$. (Is this a function?)

Find the maximum or minimum points on the graph for the following two problems.

8. $y = x^2 - 9x + 8$
9. $y = -x^2 + 9x - 8$
10. In $y = x^2 + 1$, can y ever equal zero?
11. Graph $x = y^2 - 9$.
12. Graph $x = y^2 - 6y + 9$.
13. Write a Fortran IF statement to test the coefficient of x^2 for $ax^2 + bx + c = 0$. If it is positive go to statement 30; if zero, go to statement 20; if negative, go to statement 10.
14. The following program in Basic illustrates Newton's method of finding a square root by successive approximations.

PSEUDOCODE	BASIC
Set number = 625	10 NUM = 625
Set divisor = number	20 DIV = NUM
Set quotient = number ÷ divisor	30 QUOT = NUM/DIV
⌐ DO the following WHILE quotient ≠ divisor	40 IF QUOT = DIV THEN GOTO 80
Set divisor = quotient + divisor	50 DIV = QUOT + DIV
Loop Set divisor = divisor ÷ 2	60 DIV = DIV/2
Set quotient = number ÷ divisor	65 QUOT = NUM/DIV
⌐ END of DO WHILE	70 GOTO 40
WRITE quotient,divisor,number	80 PRINT QUOT,DIV,NUM
STOP	90 END

The process begins by setting the divisor equal to the dividend:

$$\frac{1}{625)\overline{625}}$$

The loop, or repeated sequence of instructions, determines the square root by averaging the sums of quotients and divisors.

Trace each instruction in the program by writing out the activity of each arithmetic instruction until QUOTIENT equals DIVISOR. (Truncate decimal fractions, using integers only for quotients.) Next, as an exercise, write the Fortran program to find the square root of any real number. You as the programmer must keep track of the decimal positions and use the correct variable names for floating-point variables. (See page 155.) Describe the algorithm first in pseudocode, then write the Fortran program.

9-3 COMMERCIAL APPLICATIONS

The quadratic function may be used in commercial computer programming to determine relationships between prices, profits, and the production of goods.

EXAMPLE 1   Letting

P represent profit

x represent production units (such as one yard, one pound, one item, etc.)

we can demonstrate an example of a functional relationship between production units and profits in the following quadratic function:

$$f(x) = P = -10x^2 + 60x - 50$$

Assuming no negative values for x (no negative production units), we can write a table of values for P and x.

x	P
0	-50
1	0
2	30
3	40

It is evident from the table of values that to operate at a profit, at least 2 units must be produced. Continuing our table and graphing these points, we find:

x	P
4	30
5	0
6	-50

Figure 9-5 shows a graph of these points. If maximum profit is to be reached, three production units must be realized.

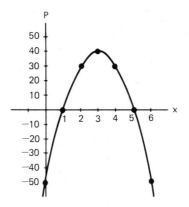

FIGURE 9-5

To find the zeros of this function, we solve for x in the equation for $f(x)$:

$$-10x^2 + 60x - 50 = 0$$
$$+10x^2 - 60x + 50 = 0$$
$$(x - 1)(x - 5) = 0$$

Then $f(x) = 0$ at $x = 1$ and $x = 5$.

Relationships also exist between costs and production.

EXAMPLE 2   Let C *represent cost in dollars.*
Let x represent production units.

Given $C = x^2 - 4x + 9$, and again assuming no negative production units, we can construct the graph shown in Fig. 9-6. The minimum cost is found at two production units.

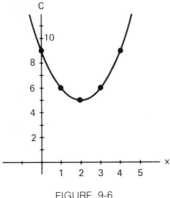

FIGURE 9-6

In cost production relationships it is important to determine minimum costs, and the point (2, 5) in Fig. 9-6 shows this minimum cost. Comparing this graph to the last (Fig. 9-5), note that the curve in Fig. 9-6 never crosses the x-axis. This means that $f(x) \neq 0$.

Relationships between profit, costs, and production units are discussed further in Chapter 11.

Exercises 9-3

1. Find the maximum profit (highest point) on the graph in Fig. 9-5.
2. Maximum profit and daily output in a garment shop are expressed in the functional relationship $P = -10x^2 + 80x - 70$.

   a. Graph the quadratic equation.
   b. Find the highest point on the graph.
   c. Find the zeros of the function.
3. If $COST = W^2 - 3W + 4$ and W represents work units, find the corresponding number of work units at the minimum cost.
4. Write Fortran statements to find the maximum profit.

_____ **9-4 LOGARITHMIC FUNCTIONS**

To study logarithmic functions, we need to combine what we have learned about relations, functions, and inverse functions with the basic laws of exponents. (Refer to page 314, Appendix B if necessary.) Consider the following:

1. If $b > 1$ and $x \in N$ (x is a member of the set of positive natural numbers; see Chapter 0, page 2), then $b^x > 1$.

EXAMPLE   Let $b = 2$, $x = 3$; then $2^3 > 1$.

2. If $0 < b < 1$, then $0 < b^x < 1$ (where $x \in N$).

EXAMPLE   Let $b = .5$, $x = 2$; then $0 < .5^2 < 1$.

3. If $b^{1/x} = a$, then $a^x = b$ (where $x \in N$).

EXAMPLE   Let $b = 8$, $x = 3$, then $8^{1/3} = 2$, and $2^3 = 8$.

Exponential functions can be easily visualized by drawing their graphs.

Let $b = 2$ in Fig. 9-7a and $b = \frac{1}{2}$ in Fig. 9-7b.

Study these two graphs and note the ordered pairs listed next to each graph. Fig. 9-7 contains graphs of functions since there is only one y associated with each x, for $f(x) = b^x$.

Notes on Fig. 9-7:

1. The domain of the function is the set of all real numbers.
2. The range of the function is the set of all positive real numbers.
3. The function $f(x) = b^x$ always passes through the point $(0, 1)$.
4. In Fig. 9-7a, the curve for $f(x) = 2^x$ goes *up* to the right and can be described as an increasing function.
5. In Fig. 9-7b, the curve for $f(x) = (\frac{1}{2})^x$ goes *down* to the right and can be described as a decreasing function.
6. Where $b > 1$, the graph has the general appearance as shown in Fig. 9-7a.
7. Where $0 < b < 1$, the graph has the general appearance as shown in Fig. 9-7b.

Powers expressed in the form of b^x, where $x \in R$, $b \in R$, and $b > 0$, $b \neq 1$, can define functions, and we can define the exponential functions for the base b as $f(x) = b^x$.

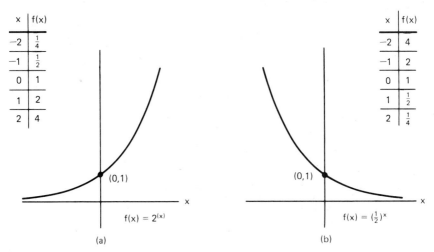

x	f(x)
−2	$\frac{1}{4}$
−1	$\frac{1}{2}$
0	1
1	2
2	4

x	f(x)
−2	4
−1	2
0	1
1	$\frac{1}{2}$
2	$\frac{1}{4}$

(0,1)

$f(x) = 2^{(x)}$

(a)

(0,1)

$f(x) = (\frac{1}{2})^x$

(b)

FIGURE 9-7

9-4.1   Inverse Functions

Substituting y for $f(x)$ in $f(x) = b^x$, we get $y = b^x$. To find the inverse of this function, we interchange x and y and write $x = b^y$. The relationship of the function to its inverse is shown in graphic form in Fig. 9-8, where $b > 1$.

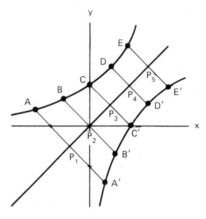

FIGURE 9-8

To produce the graph, draw the line for $y = x$. Draw $y = b^x$ and mark off points A, B, C, D, E. Draw a line from A perpendicular to the line $y = x$ and extend this line to its own length to A'. Continue this procedure for B, C, D, and E. Then draw a smooth curve connecting these points and representing $x = b^y$. (Note that $AP_1 = P_1A'$, $BP_2 = P_2B'$, etc.)

Observe that

1. The domain of the function $x = b^y$ is the set of all positive real numbers.
2. The range of the function is the set of all real numbers.

As an exercise, draw the graph for $x = b^y$, where $0 < b < 1$. Does the graph of $x = b^y$ always pass through (1, 0) when

$$y \in J,$$
x is a positive real number, and
$b > 1$ or $0 < b < 1$?

Definition 9-2

The symbolism that expresses y in terms of x for this inverse function is $y = \log_b x$ and is read "y is the logarithm of the number x to the base b." The logarithm is the exponent to which b is raised to obtain the number x.

$x = b^y$ is called the exponential form.
$y = \log_b x$ is called the logarithmic form.

One can see from this that

$$x = b^{\log_b x} \text{ and } y = \log_b b^y,$$
$$\text{since} \quad x = b^y \text{ if and only if } y = \log_b x.$$

9-4.2 Properties of Logarithms

Because logarithms can be considered as exponents, the laws governing the operations of logarithms have their basis in the laws of exponents. Restating two of these laws of exponents:

I. $b^m \times b^n = b^{m+n}$

II. $\dfrac{b^m}{b^n} = b^{m-n} = \dfrac{1}{b^{n-m}}$

A third law that we are concerned with involves roots and exponents:

III. $\sqrt[n]{b^m} = (\sqrt[n]{b})^m = b^{m/n}$

Logarithms are useful in computations involving multiplication, division, exponentiation, and roots. Let us consider the properties of logarithms for $x, y > 0$ and $x, y \in R$.

1. Applying law I, we can state that the logarithm of a product is equal to the sum of the logarithms, or $\log_b x + \log_b y = \log_b x \cdot y$.
2. Applying law II to logarithms, we can state that the logarithm of a quotient is equal to the difference of the logarithms, or $\log_b x/y = \log_b x - \log_b y$.
3. In $\sqrt[n]{x^m}$ we can write the following for logarithms: $\log_b \sqrt[n]{x^m} = m/n$ $\log_b x$, where $n \neq 0$, $m \in J$ (or, $n \neq 0$ and m and n are integers).
4. A special case of property 3 states that $\log_b (x)^m = m \log_b x$ (where $m \in J$), or that the logarithm of a power of x is equal to the product of the power times the logarithm of x. Using law III, we can now write:

$$\log_b \sqrt[n]{x^m} = \log_b x^{m/n} = m/n \log_b x$$

EXAMPLE 1 $\log_2 32 = 5$ is read, "the log of 32 to the base 2 is 5," and the $\log_2 32 = 5$ is equivalent to the exponential form $2^5 = 32$.

EXAMPLE 2 $\log_3 \frac{1}{27} = -3$, is read, "the log of $\frac{1}{27}$ to the base 3 is -3," and $\log_3 \frac{1}{27} = -3$ is equivalent to the exponential form $3^{-3} = \frac{1}{27}$.

Exercises 9-4

Given the equations below, find the second member of the ordered pairs (example: $y = 2^x$; (0, 1), (1, 2), (2, 4)).

1. $y = 3^x$; (−2,), (2,), (0,)
2. $y = -2^x$; (−1,), (0,), (1,)
3. $y = (\frac{1}{2})^x$; (−2,), (−1,), (3,)
4. $y = 10^x$; (−3,), (2,), (4,)

Graph the following functions and their corresponding inverse functions.

5. $y = 3^x$
6. $y = 2^{-x}$
7. $y = (\frac{1}{2})^x$
8. $y = (\frac{1}{3})^{-x}$

Express in exponential notation:

9. $\log_2 32 = 5$
10. $\log_{10} 100 = 2$
11. $\log_{10} 0.001 = -3$
12. $\log_{1/2} 4 = -2$
13. $\log_{16} 1 = 0$
14. $\log_5 125 = 3$

Express in logarithmic notation:

15. $4^3 = 64$
16. $8^{-1/3} = \frac{1}{2}$
17. $5^0 = 1$
18. $10^{-3} = 0.001$
19. $(\frac{1}{4})^2 = \frac{1}{16}$
20. $1000^{-1/3} = \frac{1}{10}$

Find x, y, or b in the following.

21. $\log_2 16 = y$
22. $\log_{10} 0.01 = y$
23. $\log_b 0.1 = -1$
24. $\log_5 x = 3$
25. $\log_b 32 = -5$
26. $\log_{10} \frac{1}{100} = y$
27. $\log_2 x = -3$
28. Draw the graph for $y = b^x$, where y is the set of positive real numbers, $x \in J$, and $b = 1$.

————— 9-5 LOGARITHMS TO THE BASE TEN

The values of y for the expression $y = \log_{10} x$ are called *common logarithms*.

In
$$\log_{10} 10 = \log_{10} 10^1 = 1$$
$$\log_{10} 100 = \log_{10} 10^2 = 2$$
$$\log_{10} 1 = \log_{10} 10^0 = 0$$
$$\log_{10} 0.1 = \log_{10} 10^{-1} = -1$$
$$\log_{10} 0.01 = \log_{10} 10^{-2} = -2$$
$$\log_{10} 0.001 = \log_{10} 10^{-3} = -3$$

we are interested in finding $\log_{10} x$ for each value of x.

9-5.1 Logarithmic Tables

Fig. 9-9 is taken from Appendix A, where a reference is provided for $\log_{10} x$ for each value of x, where $1 \leqslant x < 10$. Each number in the column headed x represents the first two significant digits of x.

Common Logarithms*

x	0	1	2	3	4	5	6	7	8	9
4.0	.6021	.6031	.6042	.6053	.6064	.6075	.6085	.6096	.6107	.6117
4.1	.6128	.6138	.6149	.6160	.6170,	.6180	.6191	.6201	.6212	.6222
4.2	.6232	.6243	.6253	.6263	.6274	.6284	.6294	.6304	.6314	.6325
4.3	.6335	.6345	.6355	.6365	.6375	.6385	.6395	.6405	.6415	.6425
4.4	.6435	.6444	.6454	.6464	.6474	.6484	.6493	.6503	.6513	.6522
4.5	.6532	.6542	.6551	.6561	.6571	.6580	.6590	.6599	.6609	.6618
4.6	.6628	.6637	.6646	.6656	.6665	.6675	.6684	.6693	.6702	.6712
4.7	.6721	.6730	.6739	.6749	.6758	.6767	.6776	.6785	.6794	.6803

* Log tables can be obtained by calculus with the aid of a computer. Many computer manufacturers supply *functions*, or prepared programs that aid in the solution of problems involving logarithms.

FIGURE 9-9

The numbers in the same row as x contain a third significant digit in x. The digits located in the table at the intersection of rows and columns form the logarithms of x.

EXAMPLE 1   Find the logarithm of 4.73.
Look for the row next to 4.7 and move right to the column under 3, and we find

$$\log_{10} 4.73 = 0.6749$$

We can also find

$$\log_{10} 4.08 = 0.6107$$
$$\log_{10} 4.10 = 0.6128$$

Because the numbers in the logarithm table are usually irrational, they are only approximations and do not exactly equal $\log_{10} x$ for some x. However, it is conventional to use the equal sign as shown.

EXAMPLE 2   Assume we needed to find $\log_{10} x$ for values of $x \geq 10$. Scientific notation makes it possible to represent any number $x \geq 10$ as a product of a number between 1 and 10 and a power of 10. Thus,

$$\log_{10} 32.1 = \log_{10} (3.21 \times 10^1)$$
$$= \log_{10} 3.21 + \log_{10} 10^1$$
$$= 0.5065 + 1$$
$$= 1.5065$$

The decimal portion of the logarithm is nonnegative and is called the *mantissa*. The integral part is simply the exponent of 10 when the number is written in scientific notation and is called the *characteristic*. The characteristic can be either positive or negative.

EXAMPLE 3 Find $\log_{10} 673000$

$$\begin{aligned}
\log_{10} 673000 &= \log_{10} (6.73 \times 10^5) \\
&= \log_{10} 6.73 + \log_{10} 10^5 \\
&= 0.8280 + 5 \\
&= 5.8280
\end{aligned}$$

EXAMPLE 4 Find $\log_{10} 0.00891$ (where $0 < x < 1$)

$$\begin{aligned}
\log_{10} 0.00891 &= \log_{10} (8.91 \times 10^{-3}) \\
&= \log_{10} 8.91 + \log_{10} 10^{-3} \\
&= 0.9499 + (-3)
\end{aligned}$$

However,

$$\begin{aligned}
&\quad -3.0000 \\
&\quad +0.9499 \\
\log_{10} 0.00891 &= -2.0501
\end{aligned}$$

To write the logarithm in a form in which the *decimal fraction is positive*, we let

$$\begin{aligned}
\log_{10} 0.00891 &= 0.9499 - 3 \\
&= 0.9499 + (7 - 10) \\
&= 7.9499 - 10
\end{aligned}$$

This conventional format makes it possible to subtract a multiple of 10 and to represent the logarithm in a manner whereby the decimal fraction is not a negative number. (Further, the original mantissa is retained.)

We can find a value for x in $\log_{10} x$ by reversing these procedures. For example, antilog$_{10}$ 2.7767 can be found by locating the mantissa 0.7767 in Appendix A and noting that the corresponding value for x is 5.98. Because the characteristic is 2, we multiply as follows: $5.98 \times 10^2 = 598$. Therefore, antilog$_{10}$ 2.7767 = 598, or $x = 598$.

9-5.2 Linear Interpolation

Because of space limitations, logarithm tables are usually given in three digits for x. To find the common logarithm of a four-digit number, we use a method called linear interpolation.

$$\text{Consider} \quad \frac{y_1}{y_2} = \frac{x_1}{x_2}$$

If we know three of these numbers, we can find the fourth.

Assume that we wish to find the logarithm of 5.662. We know that $5.660 < 5.662 < 5.670$. (Note that we have written the four-digit numbers 5.660 and 5.670 instead of the three-digit numbers 5.66 and 5.67.) We can now denote the 2 in 5.662 as $\frac{2}{10}$ of the distance between

5.660 and 5.670. This distance can be expressed as a ratio using logarithms. Thus, in

$$\frac{5.662 - 5.660}{5.670 - 5.660} = \frac{y}{0.7536 - 0.7528}$$

Note that

$$\frac{5.662 - 5.660}{5.670 - 5.660} = \frac{0.002}{0.010}$$

$$= \frac{0.002 \times 10^3}{0.010 \times 10^3} = \frac{2}{10}$$

Using a more conventional method, we have

$$10\left\{2\begin{cases}5.660 \\ 5.662 \\ 5.670\end{cases}\begin{matrix}0.7528 \\ 0.7536\end{matrix}\left.\begin{matrix}\\ \end{matrix}\right\}y\right\}0.0008$$

To find the difference between the logs of 5.66 and 5.67, we subtract:

$$\begin{matrix}0.7536 \\ -0.7528 \\ \hline 0.0008\end{matrix}$$

It follows that

$$\frac{2}{10} = \frac{y}{0.0008}$$

and $2/10(0.0008) = y = 0.00016$. Adding 0.00016 to 0.7528 we get 0.7530 (after rounding). This represents an approximation of the required logarithm, or $\log_{10} 5.662 = 0.7530$.

9-5.3 Sample Logarithmic Calculations

Solve the following four problems using logarithms.

1. $\dfrac{(2.37) \times (7.18)}{(4.50)}$

$$[\log_{10} (2.37) + \log_{10} (7.18)] - \log_{10} (4.50)$$
$$= [0.3747 + 0.8561 - 0.6532]$$
$$= 1.2308 - 0.6532$$

0.5776 = the logarithm of the answer. To find the antilog$_{10}$ 0.5776, we interpolate as follows:

$$11\left\{1\begin{cases}.5775 \\ .5776 \\ .5786\end{cases}\begin{matrix}3.780 \\ 3.790\end{matrix}\left.\begin{matrix}\\ \end{matrix}\right\}x\right\}0.010$$

$$\frac{1}{11} = \frac{x}{0.010}$$

$.001 = x$ and $3.780 + 0.001 = 3.781$

Thus, antilog 0.5776 = 3.781. The antilog provides a way of expressing the answer to [(2.37) × (7.18)]/(4.50) = 3.781

2. $\dfrac{(3.75)^2 \times (8.16)^{1/2}}{(7.56)^{-3}}$

    $2 \log_{10} (3.75) + \frac{1}{2} \log_{10} (8.16) - (-3) \log_{10} (7.56)$
    $= 2(0.5740) + \frac{1}{2}(0.9117) + 3(0.8785)$
    $= 1.1480 + 0.4558 + 2.6355 = 4.2393$

$$25 \left\{ 13 \left\{ \begin{array}{l} .2380 \\ .2393 \\ .2405 \end{array} \right. \quad \begin{array}{l} 1.730 \\ \\ 1.740 \end{array} \right\} \times \left. \right\} 0.010$$

    $\dfrac{13}{25} = \dfrac{x}{0.010}$

    $.005 = x$
    antilog$_{10}$ 4.2393 = 17,350

3. $(580) \times (11.2)^5 \times (3.75)^{-7}$
    $\log_{10} (580) + 5 \log_{10} (11.2) + (-7) \log_{10} (3.75)$
    $= 2.7634 + 5(1.0492) - 7(0.5740)$
    $= 2.7634 + 5.2460 - 4.0180$
    $= 8.0094 - 4.0180$
    $= 3.9914$

$$5 \left\{ 2 \left\{ \begin{array}{l} .9912 \\ .9914 \\ .9917 \end{array} \right. \quad \begin{array}{l} 9.800 \\ \\ 9.810 \end{array} \right\} \times \left. \right\} 0.010$$

    $\dfrac{2}{5} = \dfrac{x}{0.010}$

    $.004 = x$
    antilog$_{10}$ 3.9914 = 9804

4. $\dfrac{(5.77)^3 \times \sqrt[3]{(7.23)^2}}{(20.9)^{1/3}}$

    $3 \log_{10} (5.77) + \frac{2}{3} \log_{10} (7.23) - \frac{1}{3} \log_{10} (20.9)$
    $= 3(0.7612) + \frac{2}{3}(0.8591) - \frac{1}{3}(1.3201)$
    $= 2.2836 + 0.5727 - 0.4400$
    $= 2.8563 - 0.4400$
    $= 2.4163$

$$16 \left\{ 13 \left\{ \begin{array}{l} .4150 \\ .4163 \\ .4166 \end{array} \right. \quad \begin{array}{l} 2.600 \\ \\ 2.610 \end{array} \right\} \times \left. \right\} 0.010$$

    $\dfrac{13}{16} = \dfrac{x}{0.010}$

    $.008 = x$
    antilog$_{10}$ 2.4163 = 260.8

Exercises 9-5

Find each logarithm from the table in Appendix A.

1. $\log_{10} 4.25$
2. $\log_{10} 62.0$
3. $\log_{10} 7.21$
4. $\log_{10} 0.00512$
5. $\log_{10} 0.127$
6. $\log_{10} 32400$
7. $\log_{10} 295$

Find each of the following.

8. $\text{antilog}_{10} 0.5478$
9. $\text{antilog}_{10} 0.9786$
10. $\text{antilog}_{10} 2.8722$
11. $\text{antilog}_{10} 1.7709 - 2$
12. $\text{antilog}_{10} 4.6522 - 3$
13. $\text{antilog}_{10} 4.6522$
14. $\text{antilog}_{10} 0.7042$
15. $\text{antilog}_{10} 3.8887$

Interpolate and find

16. $\log_{10} 6437$
17. $\log_{10} 73.82$
18. $\log_{10} 0.1235$
19. $\log_{10} 26.41$
20. $\log_{10} 7.398$
21. $\log_{10} 534.6$
22. $\log_{10} 0.04891$
23. $\text{antilog}_{10} (6.7051 - 8)$
24. $\text{antilog}_{10} 2.9958$
25. $\text{antilog}_{10} 1.7240$

Compute each of the following using logarithms.

26. $(12.4)(7.91)$
27. $\dfrac{4.12}{1.25}$
28. $(6.41)^3$
29. $\sqrt[5]{7.61}$
30. $\dfrac{(14.8)(0.021)}{394}$
31. $\dfrac{(3.45)^3(\sqrt[2]{6.71})}{8.55}$

32. A bank pays 5.25% interest per year compounded quarterly.

   Let: P = the amount of money deposited in a savings account
   m = the number of interest periods per year
   n = the number of years the amount of money is in the bank
   r = the rate of interest
   A = the final amount (or current amount)

   The formula used to compute interest compounded quarterly is

   $$A = P \left(1 + \frac{r}{m}\right)^{mn}$$

   a. Using logarithms, find the amount to be paid a depositor whose original (and only) deposit was $200.00 made exactly 10 years ago.
   b. Write the Fortran statement(s) to compute a depositor's interest using the formula given in this question.
   c. In· 1626 Peter Minuit purchased Manhattan Island for about $24.00. If the Indians invested this money at 4% compounded semiannually, what would their current amount be?
   d. Assume the Indians desired a computer print out of A for every 10-year period since 1626. Flowchart the logic used to compute and print out A for every such 10-year period.

———————— 9-6 LOGARITHMS AND COMPUTERS

As noted earlier in this chapter, exponents and logarithms can be evaluated by computers in so-called "library" routines, or special programs written by the computer manufacturers. These special programs can be referenced by the coding shown in Fig. 9-10.*

	Function	Coding
PL/I	Exponent	EXP(X)
	Logarithm	LOG10(X)
Fortran	Exponent	EXP(X)
	Logarithm	ALOG(X)
Basic	Exponent	EXP(x)
	Logarithm	CLG(x)

FIGURE 9-10

EXAMPLE 1   Given: $a = e^2$
               Coded: $A = $ EXP(2);

After the statement $A = $ EXP(2) has been executed, A will contain the base of the natural logarithm system raised to the power of 2.

———————————
* Students should reference the manual for the computer they are using.

EXAMPLE 2   Given: $C = \log_{10} A - \log_{10} B$ (or $A \div B$)

           Coded: $C = \text{LOG10(A)} - \text{LOG10(B)};$

           or: $C = \text{LOG10(A/B)};$

The value of the common logarithm is assigned to C, where A and $B > 0$.

    Examples 1 and 2 above are *PL/I* examples. As an exercise, recode these examples in Fortran.

Exercises 9-6

1. Solve problem 32c in Exercises 9-5 with a Fortran statement that uses logarithms.

2 a. The number of bacteria in a culture is a function of time such that from one generation to the next, the number of bacteria doubles. Consider one thousand bacteria growing in a petri dish. If we wanted to find out the number of generations that would be required to produce a count of one billion (or 10^9) bacteria, we could use the formula $a = b \times 2^n$, where

        b = the original bacteria count

        a = the final bacteria count

        n = the number of generations

    Solve this problem using logarithms.

  b. If the production of each generation requires 30 minutes, how much time is needed to produce the number of generations found in part a above?

10

MATRIX ALGEBRA

It is often useful to group real numbers (or other suitable items) into a rectangular array called a matrix. Consider, for example, a situation where some items in a shoe store are to be discounted. We could enter the costs of these items into an array as follows:

Sizes

	6	7	8	9	10
shoes	20.00	20.50	21.00	22.00	23.00
sandals	8.50	9.00	9.50	10.00	10.50
boots	22.50	24.00	25.00	26.00	26.50
sneakers	4.00	4.30	4.80	5.10	5.50

If each of these items were to be reduced in price by 20 percent, each entry in the array would be multiplied by .80. For size 6 shoes, this works out as $.80 \times 20.00 = 16.00$.

Definition 10-1

A *matrix* is a rectangular array of elements (usually real numbers) displayed in brackets and often identified by upper case letters.

Examples of matrices are:

$$A = \begin{bmatrix} 1 & 4 & 7 \\ 2 & 5 & 8 \\ 3 & 6 & 9 \end{bmatrix} \qquad B = [1 \quad 2 \quad 3] \qquad C = \begin{bmatrix} 1 \\ 2 \\ 3 \end{bmatrix}$$

(a)   (b)   (c)

A matrix consisting of a single row is referred to as a *row matrix* (see (b) of the above illustration). A matrix consisting of a single column is referred to as *column matrix* (see (c) of the above illustration).

The *dimension,* or *order,* of a matrix is denoted first by its number of rows and then by its number of columns. In the above illustration, the three matrices A, B, and C are described as 3×3 (three rows by three columns) or simply "three by three," 1×3 ("one by three"), and 3×1 ("three by one"), respectively.

Matrix entries can be represented by subscripted lower-case letters. Generally, one of two formats is used.

Single Subscript Method

$$A = \begin{bmatrix} a_1 & b_1 & c_1 \\ a_2 & b_2 & c_2 \\ a_3 & b_3 & c_3 \end{bmatrix}$$

In this 3×3 matrix called A, we used a different lower-case letter for each entry in a row, but kept the same subscript denoting the row in which the element appears. Column entires are denoted by the same letter with different subscripts.

Double Subscript Method

$$A = \begin{bmatrix} a_{11} & a_{12} & a_{13} \\ a_{21} & a_{22} & a_{23} \\ a_{31} & a_{32} & a_{33} \end{bmatrix} \qquad B = \begin{bmatrix} b_{11} & b_{12} & b_{13} \\ b_{21} & b_{22} & b_{23} \end{bmatrix}$$

In the double subscript method, a single lower case letter with two subscripts is used to denote each entry. The first subscript indicates the row, and the second subscript denotes the column.

Subscripts can be represented by lower case letters, and conventionally, i is used to represent rows and j to represent columns. Thus, we can refer to the ith row or the jth column.

In much of the work in this chapter we will concentrate on square matrices, or matrices of the order $n \times m$. Although many of the ideas discussed are applicable to matrices of any order, we will apply the notion to most of those usually found in data processing situations.

Equal Matrices

Definition 10-2

Two matrices are said to be *equal* if and only if they are both of the same dimension or order and corresponding entries are equal.

$$\begin{bmatrix} 1 & 3 & 4 \\ 0 & -2 & 6 \end{bmatrix} = \begin{bmatrix} 1 & \frac{6}{2} & \frac{-12}{-3} \\ x - x & -2 & 3 \cdot 2 \end{bmatrix}$$

In general, for each *i*th row and *j*th column, $a_{ij} = b_{ij}$. This also applies to single-row and single-column matrices, where $a_{1j} = b_{1j}$ for row matrices and $a_{i1} = b_{i1}$ for column matrices. Thus,

$$[4 \quad 9 \quad 5] = [2^2 \quad 3^2 \quad \sqrt{25}]$$

$$\begin{bmatrix} 3^{-2} \\ 8^{1/3} \\ 5^2 \end{bmatrix} = \begin{bmatrix} \frac{1}{9} \\ 2 \\ 25 \end{bmatrix}$$

One of the objectives of this chapter is to provide the student with another tool for solving linear equations. Before we can use matrices in solving linear equations, we must learn how to perform certain manipulations on matrices.

10-2 MATRIX OPERATIONS

10-2.1 Transpose of a Matrix

Definition 10-3

The *transpose* of a matrix is a manipulation in which the rows are exchanged with the columns of the same matrix.

If the entries in A are A_{ij}, the entries in A transpose are A_{ji}. For example,

$$\begin{bmatrix} \sqrt{2} & 3 \\ -4 & 0 \end{bmatrix} \quad \text{transposed becomes} \quad \begin{bmatrix} \sqrt{2} & -4 \\ 3 & 0 \end{bmatrix}$$

The transpose of a matrix is denoted by a lower-case *t* as shown below. Thus,

$$\begin{bmatrix} 1 & 2 & -4 \\ 0 & 1 & 3^2 \end{bmatrix}^t = \begin{bmatrix} 1 & 0 \\ 2 & 1 \\ -4 & 3^2 \end{bmatrix}$$

10-2.2 Addition of Matrices

Definition 10-4

Matrices of the *same dimension or order* can be added, and the sums consist of the sums of the corresponding entries or elements. Thus, given

$$A = [1 \quad 2 \quad 3]$$
$$B = [-4 \quad 5 \quad 0.5]$$
$$\text{Then } A + B = [-3 \quad 7 \quad 3.5]$$

(Column matrices with the same number of entries or elements can also be added together.)

In:

$$A = \begin{bmatrix} 1 & 6 \\ 4 & -2 \end{bmatrix} \quad \text{and} \quad B = \begin{bmatrix} 14 & 0 \\ 8 & 4 \end{bmatrix}$$

$$A + B = \begin{bmatrix} 1 & 6 \\ 4 & -2 \end{bmatrix} + \begin{bmatrix} 14 & 0 \\ 8 & 4 \end{bmatrix} = \begin{bmatrix} 1+14 & 6+0 \\ 4+8 & -2+4 \end{bmatrix} = \begin{bmatrix} 15 & 6 \\ 12 & 2 \end{bmatrix}$$

Note that the sums are placed in a third matrix of the same dimension. In a general way we can write:

$$\begin{bmatrix} a_1 & b_1 \\ a_2 & b_2 \\ a_3 & b_3 \end{bmatrix} + \begin{bmatrix} x_1 & y_1 \\ x_2 & y_2 \\ x_3 & y_3 \end{bmatrix} = \begin{bmatrix} a_1 + x_1 & b_1 + y_1 \\ a_2 + x_2 & b_2 + y_2 \\ a_3 + x_3 & b_3 + y_3 \end{bmatrix}$$

The addition of matrices of different dimensions is not defined.

10-2.3 Multiplication by a Constant

Definition 10-5

A matrix can be multiplied by a constant or a variable. The product of the matrix and the constant or variable is another matrix whose entries contain the products of the constant or variable, and corresponding entries of the original matrix.

EXAMPLES:

$$5 \times \begin{bmatrix} \frac{1}{5} \\ 2 \\ 0.5 \end{bmatrix} = \begin{bmatrix} 1 \\ 10 \\ 2.5 \end{bmatrix}$$

$$3 \times \begin{bmatrix} 4 & \frac{1}{3} \\ 6 & -1 \end{bmatrix} = \begin{bmatrix} 12 & 1 \\ 18 & -3 \end{bmatrix}$$

In general, we can write

$$k \times \begin{bmatrix} a_1 & b_1 & c_1 \\ a_2 & b_2 & c_2 \\ a_3 & b_3 & c_3 \end{bmatrix} = \begin{bmatrix} ka_1 & kb_1 & kc_1 \\ ka_2 & kb_2 & kc_2 \\ ka_3 & kb_3 & kc_3 \end{bmatrix}$$

Let

$$A = \begin{bmatrix} 3 & 7 \\ -1 & .5 \end{bmatrix}$$

Suppose it is desirable to multiply matrix A by the constant -1. Then

$$-1 \times \begin{bmatrix} 3 & 7 \\ -1 & .5 \end{bmatrix} = \begin{bmatrix} -3 & -7 \\ 1 & -.5 \end{bmatrix}$$

and we can write $-A = (-1) \cdot A$.

10-2.4   Subtraction of Matrices

The negative of a matrix is formed by replacing each entry by its *additive inverse*. (See page 5.)
Thus, if

$$A = \begin{bmatrix} 3 & -2 & 1 \\ 4 & 8 & -5 \end{bmatrix}$$

then

$$-A = \begin{bmatrix} -3 & 2 & -1 \\ -4 & -8 & 5 \end{bmatrix}$$

Definition 10-6

The sum of a matrix and the negative of another matrix of the same dimension is represented as $A + (-B)$ or $A - B$. The resulting matrix contains entries representing the differences between corresponding entries.

In general, we can write

$$\begin{bmatrix} a_{11} & a_{12} \\ a_{21} & a_{22} \end{bmatrix} - \begin{bmatrix} b_{11} & b_{12} \\ b_{21} & b_{22} \end{bmatrix} = \begin{bmatrix} a_{11} & a_{12} \\ a_{21} & a_{22} \end{bmatrix} + \begin{bmatrix} -b_{11} & -b_{12} \\ -b_{21} & -b_{22} \end{bmatrix}$$

If

$$A = \begin{bmatrix} 1 & 3 \\ 0 & 4 \end{bmatrix} \text{ and } B = \begin{bmatrix} -8 & 11 \\ 9 & 2 \end{bmatrix}$$

then

$$A + (-B) = A - B = \begin{bmatrix} 9 & -8 \\ -9 & 2 \end{bmatrix}$$

Exercises 10-1

1. Write a matrix equal to

$$\begin{bmatrix} 1 & 0 & 2^2 \\ 4 & 1 & \frac{1}{2} \\ 3 & .06 & -4 \end{bmatrix}$$

2. Name the order of

a. $\begin{bmatrix} a_{11} & a_{12} \\ a_{21} & a_{22} \\ a_{31} & a_{33} \end{bmatrix}$ b. $\begin{bmatrix} a \\ b \\ c \end{bmatrix}$

c. $[-4 \quad 3 \quad .05]$

3. Find the transpose of

a. $\begin{bmatrix} 1 & -2 \\ 3 & \frac{6}{7} \end{bmatrix}$ b. $\begin{bmatrix} 2 & 1 \\ 3 & -4 \\ 0 & 6 \end{bmatrix}$

c. $[1 \quad 2 \quad 3]$ d. $\begin{bmatrix} 1 & 2 & 3 & 4 \\ 5 & 6 & 7 & 8 \\ 9 & 10 & 11 & 12 \\ 13 & 14 & 15 & 16 \end{bmatrix}$

4. Add:

a. $\begin{bmatrix} 1 & 2 & 6 \\ 4 & 0 & \frac{1}{2} \end{bmatrix} + \begin{bmatrix} 3 & -2 & 8 \\ -1 & -1 & -2 \end{bmatrix}$

b. $\begin{bmatrix} 1 & -\frac{1}{3} \\ 2 & \frac{7}{6} \\ 3 & 0 \\ 4 & -1 \end{bmatrix} + \begin{bmatrix} 3 & -\frac{2}{3} \\ -6 & \frac{5}{6} \\ 4 & 1 \\ \frac{1}{2} & 0 \end{bmatrix}$

c. Let $A = \begin{bmatrix} 1 & -3 \\ 6 & 2 \end{bmatrix}$ and $B = \begin{bmatrix} 7 & 14 \\ 0 & -1 \end{bmatrix}$

   Is addition of matrices commutative, that is, does $A + B = B + A$?

5. Let a matrix whose entries or elements are all equal to zero be called a zero matrix.

   a. Does the zero matrix function as an identity matrix for addition?
   b. Give an example to determine whether or not $A + 0 = 0 + A = A$.

6. Find the negative of

a. $\begin{bmatrix} 1 & -3 & 6 \\ 4 & 7 & -2 \\ -3 & -1 & 6 \end{bmatrix}$ b. $\begin{bmatrix} 3 & 7 \\ -6 & 9 \\ 4 & -2 \\ \frac{1}{3} & -1 \end{bmatrix}$

7. Subtract:

a. $\begin{bmatrix} 3 & 4 \\ 2 & 1 \end{bmatrix} - \begin{bmatrix} 2 & 4 \\ 1 & 3 \end{bmatrix}$
b. $\begin{bmatrix} 4 \\ 0 \\ 1 \end{bmatrix} - \begin{bmatrix} 3 \\ 6 \\ 0 \end{bmatrix}$

8. Solve for C.

a. $C + \begin{bmatrix} 1 & -3 \\ 2 & 0 \end{bmatrix} = \begin{bmatrix} 1 & 2 \\ 1 & 4 \end{bmatrix}$

b. $C - \begin{bmatrix} 4 & 8 & -11 \\ 15 & \frac{1}{3} & 4 \end{bmatrix} = \begin{bmatrix} 2 & 6 & -9 \\ 13 & 1 & 0 \end{bmatrix}$

c. $C + \begin{bmatrix} 2 & 4 & 7 \\ 3 & -1 & 9 \\ 2 & 0 & 1 \end{bmatrix}^t = \begin{bmatrix} 0 & 0 & 0 \\ 0 & 0 & 0 \\ 0 & 0 & 0 \end{bmatrix}$

9. Given $A = \begin{bmatrix} -3 & 6 \\ -2 & 1 \end{bmatrix}$

$B = \begin{bmatrix} 1 & 0 \\ 3 & 1 \end{bmatrix}$

Find C in $A + B = C$ and show that $B = C - A$.

10. Multiply:

a. $-3 \times \begin{bmatrix} -4 & \frac{1}{3} \\ 2 & 0 \end{bmatrix}$

b. $y \times [a \quad -1 \quad \frac{2}{3}]$

_____ 10-3 **MULTIPLICATION OF MATRICES**

The product of two matrices A and B is a resulting matrix with entries determined as follows:
if

$$A = \begin{bmatrix} a_{11} & a_{12} \\ a_{21} & a_{22} \end{bmatrix}$$

and

$$B = \begin{bmatrix} b_{11} & b_{12} \\ b_{21} & b_{22} \end{bmatrix}$$

then

$$AB = \begin{bmatrix} (a_{11}b_{11} + a_{12}b_{21}) & (a_{11}b_{12} + a_{12}b_{22}) \\ (a_{21}b_{11} + a_{22}b_{21}) & (a_{21}b_{12} + a_{22}b_{22}) \end{bmatrix}$$

or

$$AB = C, \text{ where } C = \begin{bmatrix} c_{11} & c_{12} \\ c_{21} & c_{22} \end{bmatrix}$$

c_{11} is determined by multiplying a_{11} and b_{11} and adding this product to the product of a_{12} and b_{21}.

such that

$$c_{11} = a_{11}b_{11} + a_{12}b_{21}$$
$$c_{21} = a_{21}b_{11} + a_{22}b_{21}$$
$$c_{12} = a_{11}b_{12} + a_{12}b_{22}$$
$$c_{22} = a_{21}b_{12} + a_{22}b_{22}$$

It is often useful to describe the dimensions of a matrix using variables. Thus, in $A_{m \times n}$, the matrix A contains m rows and n columns.

Definition 10-7

The product of the matrices $A_{m \times n}$ and $B_{n \times p}$ is the matrix $(AB)_{m \times p}$, with entries developed as follows:

The entry c_{ij} in the product of A and B is determined by multiplying the first entry in the ith row of matrix A by the first entry in the jth column of matrix B and adding this product to the product formed by multiplying the second element in the ith row of A with second element in the jth column of B, etc.

In a 3×3 matrix this works out as follows. Let

$$A = \begin{bmatrix} 1 & 2 & -3 \\ 4 & 0 & 1 \\ 6 & \frac{1}{2} & -1 \end{bmatrix} \qquad B = \begin{bmatrix} 0 & 1 & 1 \\ \frac{1}{2} & -2 & -4 \\ 1 & 2 & 1 \end{bmatrix}$$

Then

$$AB = \begin{bmatrix} 0+1-3 & 1-4-6 & 1-8-3 \\ 0+0+1 & 4+0+2 & 4+0+1 \\ 0+\frac{1}{4}-1 & 6-1-2 & 6-2-1 \end{bmatrix} = \begin{bmatrix} -2 & -9 & -10 \\ 1 & 6 & 5 \\ -\frac{3}{4} & 3 & 3 \end{bmatrix}$$

Unlike matrix addition, matrix multiplication is not limited to matrices having the same dimensions. Where the first matrix has the same number of columns as the second matrix has rows, we say that the two matrices are conformable for multiplication. For example,

$$\begin{bmatrix} 1 & 2 & -4 \\ 3 & 7 & 5 \end{bmatrix} \times \begin{bmatrix} 3 & -1 \\ 1 & 2 \\ 2 & 3 \end{bmatrix} = \begin{bmatrix} 3+2+(-8) & -1+ \ 4+(-12) \\ 9+7+ \ \ 10 & -3+14+ \ \ 15 \end{bmatrix}$$

$$= \begin{bmatrix} -3 & -9 \\ 26 & 26 \end{bmatrix}$$

Notice that our resulting matrix is 2×2. In general, we can say that the outer two dimension numbers of the original matrices form the dimensions of the resulting matrix. We can illustrate our last example as follows:

$2 \times 3 \qquad 3 \times 2$

Indicates the order of the resulting matrix.

If equal, matrices are conformable.

If our original matrices were 3 × 2 and 2 × 3, they would be conformable for multiplication. In

$$3 \times 2 \qquad 2 \times 3$$

our resulting matrix would have the dimensions of 3 × 3. An example of two matrices *not* conformable for multiplication is

$$\begin{bmatrix} 1 & 2 & 3 \\ 4 & 5 & 6 \end{bmatrix} \cdot \begin{bmatrix} 7 \\ 8 \end{bmatrix}$$

Since the first matrix does not have the same number of columns as the second matrix has rows, these two matrices are not conformable for multiplication. (The division of matrices is not defined in this section of the chapter. The multiplicative inverse of a square matrix is discussed on page 226.)

Exercises 10-2

1. Multiply.

a. $3 \times \begin{bmatrix} 4 & 6 \\ -2 & \frac{1}{3} \end{bmatrix}$

b. $.02 \times \begin{bmatrix} 10.32 & 31.8 & 6.00 \\ 64.71 & 4.09 & -4.13 \end{bmatrix}$

2. Division in matrix algebra has not been defined, but we can multiply a matrix by a constant rational number. Accordingly, find

$$\frac{1}{2} \times \begin{bmatrix} 6 & -3 \\ 2 & 0 \\ 4 & 8 \end{bmatrix}$$

3. The *XYZ* Company has a chain of motels across the state. In order to meet the rising costs of living, a 5% salary increase has been given each of sixteen managers. If

4 managers earn $ 9,500 yearly
6 managers earn $10,000 yearly
3 managers earn $11,000 yearly
2 managers earn $11,500 yearly
1 manager earns $13,000 yearly

display their respective salary increases in matrix form.

4. Multiply.

a. $\begin{bmatrix} 4 & 7 \\ 3 & 5 \end{bmatrix} \cdot \begin{bmatrix} 1 & -2 \\ 0 & 4 \end{bmatrix}$

b. $\begin{bmatrix} 3 & 2 & -4 \\ 0 & 5 & 6 \\ 1 & 2 & 2 \end{bmatrix} \cdot \begin{bmatrix} 1 & -1 & 1 \\ -1 & 1 & -1 \\ 1 & -1 & 1 \end{bmatrix}$

c. $\begin{bmatrix} 2 & 0 & 4 \\ 1 & 1 & -1 \end{bmatrix} \cdot \begin{bmatrix} 1 & 3 \\ 0 & -2 \\ 4 & 1 \end{bmatrix}$

5. a. Let $A = \begin{bmatrix} 1 & -1 \\ 2 & -2 \\ 3 & -3 \end{bmatrix}$

   $B = \begin{bmatrix} 1 & 2 & 3 \\ -1 & -2 & -3 \end{bmatrix}$

   Does $AB = BA$? Is the multiplication of matrices (generally) commutative?

   b. Let $A = \begin{bmatrix} 1 & 0 \\ 2 & 4 \end{bmatrix}$

   $B = \begin{bmatrix} 1 & 4 \\ 2 & 1 \\ 0 & -2 \end{bmatrix}$

   Does $AB = BA$?

   c. $\begin{bmatrix} 1 & -2 & 4 \end{bmatrix} \cdot \begin{bmatrix} 2 \\ -1 \\ 3 \end{bmatrix} = ?$

6. Let $A = \begin{bmatrix} 3 & 2 \\ 1 & -1 \end{bmatrix}$

   $B = \begin{bmatrix} 2 & 1 \\ -1 & 1 \end{bmatrix}$

   Find a. AB
   b. BA^t
   c. B^2A or $B \cdot B \cdot A$
   d. A^2B or $A \cdot A \cdot B$
   e. $B^t \cdot A^t$
   f. $(AB)^t$
   g. Does $B^t \cdot A^t = (BA)^t$?
      If not, what is $(BA)^t$?

7. Given $A = \begin{bmatrix} a_{11} & a_{12} \\ a_{21} & a_{22} \end{bmatrix}$

   $B = \begin{bmatrix} b_{11} & b_{12} \\ b_{21} & b_{22} \end{bmatrix}$

   $C = \begin{bmatrix} c_{11} & c_{12} \\ c_{21} & c_{22} \end{bmatrix}$

   Does a. $A \cdot (B + C) = AB + AC$
   b. $A \cdot (B + C) = (A + B)(A + C)$
   c. $A \cdot (B \cdot C) = (A \cdot B) \cdot C$ (Is the multiplication of matrices associative?)

   Perform the operations enclosed in parentheses first.

10-4 DETERMINANTS

Each square matrix (or $n \times n$ matrix) containing real number entries has associated with it another real number called the determinant.

Definition 10-8

The *determinant* of the 2×2 square matrix A given by

$$\begin{vmatrix} a_{11} & a_{12} \\ a_{21} & a_{22} \end{vmatrix}$$

is the real number $a_{11} \cdot a_{22} - a_{12} \cdot a_{21}$, where $a_{11}, a_{12}, a_{21}, a_{22} \in R$.

The symbol for the determinant of A is $|A|$. Let

$$A = \begin{bmatrix} 2 & -1 \\ 3 & 4 \end{bmatrix}$$

Then $|A| = \begin{vmatrix} 2 & -1 \\ 3 & 4 \end{vmatrix} = 2 \cdot 4 - (-1) \cdot 3 = 11$

Determinants are a useful tool in solving simultaneous equations. Consider the following general format for a linear system in two variables, x and y. Let a, b, d, and e represent the coefficients on these variables. Let c and f represent values for these equations.

$$ax + by = c \tag{10-1}$$
$$dx + ey = f \tag{10-2}$$

To solve these two equations for y:

Step 1. Multiply (10-1) by $-d$.

$$-dax + (-dby) = -dc \tag{10-3}$$

Step 2. Multiply (10-2) by a.

$$adx + aey = af \tag{10-4}$$

Step 3. Add (10-3) to (10-4).

$$aey - dby = af - dc$$

Step 4. Factor out y.

$$y(ae - db) = af - dc$$

Step 5. Solve for y.

$$y = \frac{af - dc}{ae - db}, \text{ provided } (a \cdot e - d \cdot b) \neq 0$$

In a similar manner, we can solve for x:

Step 1. Multiply (10-1) by e.

$$eax + eby = ec \tag{10-5}$$

Step 2. Multiply (10-2) by $-b$.

$$-bdx - bey = -bf \qquad (10\text{-}6)$$

Step 3. Add (10-5) to (10-6).

$$eax - bdx = ec - bf$$

Step 4. Factor out x.

$$x(ea - bd) = ec - bf$$

Step 5. Solve for x.

$$x = \frac{ec - bf}{ea - bd}, \text{ provided } (a \cdot e - b \cdot d) \neq 0$$

If the coefficients of x and y are not zero, we can find values for x and y in terms of the values expressed on the right-hand side of the equals sign. (These values are typically real numbers.)

Recall that the determinant of matrix $\begin{bmatrix} a_{11} & a_{12} \\ a_{21} & a_{22} \end{bmatrix}$ is

$$\begin{vmatrix} a_{11} & a_{12} \\ a_{21} & a_{22} \end{vmatrix} = a_{11} \cdot a_{22} - a_{12} \cdot a_{21}$$

Using the coefficients from (10-1) and (10-2), let $a_{11} = a$, $a_{12} = b$, $a_{21} = d$, and $a_{22} = e$. Then we have the determinant

$$A = \begin{vmatrix} a & b \\ d & e \end{vmatrix} = ae - bd$$

Further, we can also get

$$B = \begin{vmatrix} c & b \\ f & e \end{vmatrix} = ce - bf$$

$$C = \begin{vmatrix} a & c \\ d & f \end{vmatrix} = af - cd$$

Then: The denominators for x and y in

$$x = \frac{ec - bf}{ae - bd}, \quad y = \frac{af - dc}{ae - db} \qquad (10\text{-}7)$$

are equal to the determinant of A and are the coefficients on x and y in the equations (10-1) and (10-2).

The determinant for B is equal to the numerator for x and the determinant for C is equal to the numerator for y. Further, the entries in the three determinants are the coefficients of the variables x and y and the values of these equations.

A system of two linear equations does not always have unique solutions. As stated in Chapter 8, some systems are dependent or inconsistent. The use of determinants makes it a simple matter to examine systems of equations for these possibilities. Consider $x = \dfrac{|B|}{|A|}$, and $y = \dfrac{|C|}{|A|}$.

In

$$\left.\begin{array}{l} 3x - y = 2 \\ -x + \tfrac{1}{3}y = -\tfrac{2}{3} \end{array}\right\} \text{ a dependent system of equations,}$$

we can express x in terms of determinants as

$$x = \frac{\begin{vmatrix} c & b \\ f & e \end{vmatrix}}{\begin{vmatrix} a & b \\ d & e \end{vmatrix}} = \frac{\begin{vmatrix} 2 & -1 \\ -\tfrac{2}{3} & \tfrac{1}{3} \end{vmatrix}}{\begin{vmatrix} 3 & -1 \\ -1 & \tfrac{1}{3} \end{vmatrix}}$$

Or

$$x = \frac{\tfrac{2}{3} - (+\tfrac{2}{3})}{1 - 1} = \frac{0}{0}$$

This example illustrates that when a system of linear equations is dependent, the determinant of the coefficients on x and y [the denominators in equations (10-7)] is zero, and we need not go any further in looking for a solution.

The following system contains inconsistent equations.

$$x - y = 2$$
$$2x - 2y = 3$$

Examining the determinant of the coefficients on x and y, we find

$$|A| = \begin{vmatrix} 1 & -1 \\ 2 & -2 \end{vmatrix} = -2 - (-2) = 0$$

and therefore no unique solution is possible. (The denominator would be zero.)

Consider the next system of equations:

$$x + 2y = 4$$
$$3x + y = 8$$

Since

$$\begin{vmatrix} 1 & 2 \\ 3 & 1 \end{vmatrix} = 1 - 6 = -5 = |A|$$

the determinant of the coefficients on x and y is not zero, and we can continue to solve for x and y, as follows:

$$x = \frac{\begin{vmatrix} 4 & 2 \\ 8 & 1 \end{vmatrix}}{\begin{vmatrix} 1 & 2 \\ 3 & 1 \end{vmatrix}} = \frac{4 - 16}{1 - 6} = \frac{12}{5}$$

$$y = \frac{\begin{vmatrix} 1 & 4 \\ 3 & 8 \end{vmatrix}}{\begin{vmatrix} 1 & 2 \\ 3 & 1 \end{vmatrix}} = \frac{8 - 12}{1 - 6} = \frac{4}{5}$$

As an exercise, set up the matrices and find the determinants for the *dependent* system of equations found on page 183 of Chapter 8.

10-4.1 The Three-by-Three Matrix

Using the double subscript method, we can write the following 3×3 matrix:

$$A = \begin{bmatrix} a_{11} & a_{12} & a_{13} \\ a_{21} & a_{22} & a_{23} \\ a_{31} & a_{32} & a_{33} \end{bmatrix}$$

The determinant of this matrix can be evaluated by the method of expansion by minors. A minor of an element or entry in a determinant is the determinant formed by deleting the row and column in which the entry appears. Thus, the minor of a_{11} is formed by deleting the first row and first column.

$$\begin{vmatrix} a_{11} & a_{12} & a_{13} \\ a_{21} & a_{22} & a_{23} \\ a_{31} & a_{32} & a_{33} \end{vmatrix} = \begin{vmatrix} a_{22} & a_{23} \\ a_{32} & a_{33} \end{vmatrix} = \text{minor of } a_{11}.$$

The minor of a_{31} is $\begin{vmatrix} a_{12} & a_{13} \\ a_{22} & a_{23} \end{vmatrix}$, the minor of a_{22} is $\begin{vmatrix} a_{11} & a_{13} \\ a_{31} & a_{33} \end{vmatrix}$, and so on.

Definition 10-9

The *minor* of an element a_{ij} in a determinant is the determinant that remains after the *i*th row and *j*th column have been deleted.

Definition 10-10

The *cofactor* of an element in a square matrix is the minor of that element multiplied by -1 or $+1$. If the sum of the digits in the subscript is an even number, multiply the minor by $+1$. If the sum of the digits in the subscript is an odd number, multiply the minor by -1.

EXAMPLE 1   Refer to Matrix A, above. The cofactor of a_{13} is

$$\begin{vmatrix} a_{21} & a_{22} \\ a_{31} & a_{32} \end{vmatrix}$$

because a_{13} contains the subscripts 1 and 3 which, when added, yield an even number $(1 + 3 = 4)$.

EXAMPLE 2   The cofactor of a_{12} is

$$-\begin{vmatrix} a_{21} & a_{23} \\ a_{31} & a_{33} \end{vmatrix}$$

or the negative of the minor of a_{12}, because $1 + 2$ yields an odd number. The cofactor is therefore a minor with a $+$ or $-$ sign. Generally, we can state that for an $n \times n$ matrix, the following pattern for alternating signs of minors holds true.

$$\begin{vmatrix} + & - & + & \cdot & \cdot \\ - & + & - & \cdot & \cdot \\ + & - & + & \cdot & \cdot \\ \cdot & \cdot & \cdot & \cdot & \cdot \\ \cdot & \cdot & \cdot & \cdot & \cdot \end{vmatrix}$$

Evaluating the determinant of a 3×3 matrix can now be described using cofactors and an expansion by minors.

Definition 10-11

The value of the determinant of a 3×3 square matrix is the sum of the products formed by multiplying all of the entries in a single row or a single column by their cofactors (or minors with a prefixed sign).

The determinant of a square matrix is expanded on the row or column chosen. Given

$$\begin{vmatrix} a_{11} & a_{12} & a_{13} \\ a_{21} & a_{22} & a_{23} \\ a_{31} & a_{32} & a_{33} \end{vmatrix}$$

if we choose to expand on the first column to find the determinant, we obtain

$$(A) \quad a_{11}\left(+\begin{vmatrix} a_{22} & a_{23} \\ a_{32} & a_{33} \end{vmatrix}\right) + a_{21}\left(-\begin{vmatrix} a_{12} & a_{13} \\ a_{32} & a_{33} \end{vmatrix}\right) + a_{31}\left(+\begin{vmatrix} a_{12} & a_{13} \\ a_{22} & a_{23} \end{vmatrix}\right)$$

If we choose to expand on the second row, we obtain

$$(B) \quad a_{21}\left(-\begin{vmatrix} a_{12} & a_{13} \\ a_{32} & a_{33} \end{vmatrix}\right) + a_{22}\left(+\begin{vmatrix} a_{11} & a_{13} \\ a_{31} & a_{33} \end{vmatrix}\right) + a_{23}\left(-\begin{vmatrix} a_{11} & a_{12} \\ a_{31} & a_{32} \end{vmatrix}\right)$$

As an exercise, show that (A) and (B) yield the same result.
Given

$$A = \begin{bmatrix} 3 & 6 & -2 \\ 1 & 4 & 5 \\ -2 & 0 & 1 \end{bmatrix}$$

find the determinant for A by expansion on the second column. (We chose the second column because the zero in the third row will simplify our calculations. For the same reason, we could have evaluated this determinant using the third row.)

Expanding yields

$$6\left(-\begin{vmatrix} 1 & 5 \\ -2 & 1 \end{vmatrix}\right) + 4\left(+\begin{vmatrix} 3 & -2 \\ -2 & 1 \end{vmatrix}\right) + 0\left(-\begin{vmatrix} 3 & -2 \\ 1 & 5 \end{vmatrix}\right)$$

$$6(-(1 + 10)) + 4(3 - 4) + 0$$
$$-66 + (-4)$$

Therefore the value of the determinant is -70

Exercises 10-3

1. Find the values of the following determinants

   a. $\begin{vmatrix} 3 & 1 \\ 4 & 6 \end{vmatrix}$

   b. $\begin{vmatrix} \frac{1}{2} & -3 \\ -2 & 6 \end{vmatrix}$

   c. $\begin{vmatrix} .75 & -2 \\ .5 & -1 \end{vmatrix}$

   d. $\begin{vmatrix} m & r \\ -s & 0 \end{vmatrix}$

2. Find the matrix for the following expressions.
   a. $(2) \cdot (-3) - (4) \cdot (2)$
   b. $(-1) \cdot (0) - (2) \cdot (-1)$
   c. $mn - pq$
   d. $(a) \cdot (b) + cd = (a) \cdot (b) - (-c) \cdot (+d)$

3. Solve the following using determinants. First determine if a unique solution exists.

   a. $\begin{cases} y = 4 + 2x \\ x + 3y = 7 \end{cases}$

   b. $\begin{cases} \dfrac{3y + x}{2} = 4 \\ x - y = 7 \end{cases}$

   c. $\begin{cases} 3x + \frac{1}{2}y = 3 \\ -6x - 3y = -6 \end{cases}$

   d. $\begin{cases} 2x + 3y = 4 \\ y - 3x = 7 \end{cases}$

   e. $\begin{cases} .5x - .75y = .25 \\ x - 1.5y = .3 \end{cases}$

   f. $\begin{cases} 3y = 2 \\ 3x + y = 7 \end{cases}$

4. Find the determinant for the following by expanding upon any row or column.

a. $\begin{bmatrix} 1 & 0 & 0 \\ -1 & 2 & 1 \\ 3 & -2 & 1 \end{bmatrix}$

b. $\begin{bmatrix} 0 & 1 & 2 \\ 1 & 0 & 3 \\ 2 & -1 & 0 \end{bmatrix}$

5. Solve for a in

a. $\begin{vmatrix} a & 2 \\ -1 & 1 \end{vmatrix} = 3$

b. $\begin{vmatrix} 1 & 0 & a \\ -2 & a & -1 \\ 0 & 1 & 1 \end{vmatrix} = 14$

_____ **10-5 THE IDENTITY MATRIX**

Field properties include identity elements, and matrix algebra includes the identity matrix for multiplication, usually denoted by I. For 2 × 2 matrices the identity matrix is

$$I = \begin{bmatrix} 1 & 0 \\ 0 & 1 \end{bmatrix}$$

For 3 × 3 matrices the identity matrix is

$$I = \begin{bmatrix} 1 & 0 & 0 \\ 0 & 1 & 0 \\ 0 & 0 & 1 \end{bmatrix}$$

The 1's appear on a line known as the major or main diagonal.

Definition 10-12

The *identity matrix* for multiplication is the matrix denoted by I such that $A \cdot I = I \cdot A = A$. A demonstration of the proof is left to the student.

10-6 MULTIPLICATIVE
_____ **INVERSE OF SQUARE MATRICES**

Field properties include the inverse property for a real number $1/a$ such that $a \times a^{-1} = 1$ or $a \times 1/a = 1$ ($a \neq 0$). Matrix algebra likewise includes the inverse property that for a square matrix A there exists an inverse matrix A denoted by A^{-1} (if the determinant of $A \neq 0$).

Definition 10-13

The *multiplicative inverse* of a square matrix is the matrix, denoted by A^{-1}, such that $A \cdot A^{-1} = A^{-1} \cdot A = I$ whenever $|A| \neq 0$.

10-6.1 Two-by-Two Matrices

$$\begin{bmatrix} a_{11} & a_{12} \\ a_{21} & a_{22} \end{bmatrix}$$

To write the inverse of a 2×2 square matrix:

Step 1. Find the value of the determinant of the matrix. (If the value of the determinant $= 0$, no further work is needed.)

Step 2. Interchange the entries on the main diagonal.

Step 3. Change a_{12} and a_{21} to their additive inverse.

Step 4. Multiply the resulting matrix by the multiplicative inverse of the determinant.

These rules are helpful in computer programming.

EXAMPLE 1   Let

$$A = \begin{vmatrix} 2 & 5 \\ 1 & -2 \end{vmatrix}$$

The steps to be followed are illustrated below.

Step 1. The determinant of A is $(-4) - (5) = -9$

Step 2. $\begin{vmatrix} -2 & 5 \\ 1 & 2 \end{vmatrix}$

Step 3. $\begin{vmatrix} -2 & -5 \\ -1 & 2 \end{vmatrix}$

Step 4. $-\frac{1}{9} \times \begin{bmatrix} -2 & -5 \\ -1 & 2 \end{bmatrix} = \begin{bmatrix} \frac{2}{9} & \frac{5}{9} \\ \frac{1}{9} & -\frac{2}{9} \end{bmatrix} = A^{-1}$

We observed in Exercise 10-2, Problem 5a that the multiplication of matrices is not *generally* commutative. As an exercise, show that $A \cdot A^{-1} = A^{-1} \cdot A = I$

EXAMPLE 2   In

$$\begin{bmatrix} 4 & 10 \\ 2 & 5 \end{bmatrix}$$

the determinant $= 0$; therefore, no inverse of this matrix exists.

10-6.2 Three-by-Three Matrices

To write the inverse of a 3 × 3 square matrix A:

Step 1. Find the value of the determinant of A. (If the determinant of A = 0, no further work is needed.)

Step 2. Find the cofactors of the entries in A and replace each entry with its cofactor.

Step 3. Find the transpose of the new matrix.

Step 4. Multiply the transpose by the multiplicative inverse of the determinant of A.

EXAMPLE 3   Let

$$A = \begin{bmatrix} 1 & 1 & -2 \\ 2 & 0 & 1 \\ 1 & -1 & 1 \end{bmatrix}$$

Step 1. Expanding on the first column, the value of the determinant of A is

$$1\left(+\begin{vmatrix} 0 & 1 \\ -1 & 1 \end{vmatrix}\right) + 2\left(-\begin{vmatrix} 1 & -2 \\ -1 & 1 \end{vmatrix}\right) + 1\left(+\begin{vmatrix} 1 & -2 \\ 0 & 1 \end{vmatrix}\right)$$

$$= 1(1) + 2(-(-1)) + 1(1) = 4$$

Since this value does not equal zero, continue on.

Step 2. Find cofactors of each entry and replace entries by cofactors.

$$\begin{bmatrix} 1 & -1 & -2 \\ 1 & 3 & 2 \\ 1 & -5 & -2 \end{bmatrix}$$

Step 3. Transpose

$$\begin{bmatrix} 1 & 1 & 1 \\ -1 & 3 & -5 \\ -2 & 2 & -2 \end{bmatrix}$$

Step 4.

$$A^{-1} = \tfrac{1}{4}\begin{bmatrix} 1 & 1 & 1 \\ -1 & 3 & -5 \\ -2 & 2 & -2 \end{bmatrix}$$

$A^{-1} \cdot A = I$; therefore, we can check our results as follows:

$$\tfrac{1}{4} \cdot \begin{bmatrix} 1 & 1 & 1 \\ -1 & 3 & -5 \\ -2 & 2 & -2 \end{bmatrix} \cdot \begin{bmatrix} 1 & 1 & -2 \\ 2 & 0 & 1 \\ 1 & -1 & 1 \end{bmatrix} = \tfrac{1}{4}\begin{bmatrix} 4 & 0 & 0 \\ 0 & 4 & 0 \\ 0 & 0 & 4 \end{bmatrix} = \begin{bmatrix} 1 & 0 & 0 \\ 0 & 1 & 0 \\ 0 & 0 & 1 \end{bmatrix}$$

Exercises 10-4

1. Find the inverse if it exists.

   a. $\begin{bmatrix} 2 & 5 \\ -1 & 1 \end{bmatrix}$

   b. $\begin{bmatrix} 1 & 3 \\ 2 & 6 \end{bmatrix}$

   c. $\begin{bmatrix} x & 3 \\ 0 & 2 \end{bmatrix}$ $(x \neq 0)$

2. If $A = \begin{bmatrix} 1 & 2 \\ 3 & 4 \end{bmatrix}$

   a. Show that $A \cdot A^{-1} = I = A^{-1} \cdot A$.
   b. Is $(A^t)^{-1} = (A^{-1})^t$?

3. Find the inverse if it exists.

   a. $\begin{bmatrix} 1 & 2 & -1 \\ 0 & 2 & 3 \\ 1 & 0 & 1 \end{bmatrix}$

   b. $\begin{bmatrix} 2 & -1 & 2 \\ 1 & -1 & 4 \\ 2 & 0 & 0 \end{bmatrix}$

   c. $\begin{bmatrix} 3 & -1 & 2 \\ 1 & -1 & 0 \\ 2 & 0 & 1 \end{bmatrix}$

4. If $A = \begin{bmatrix} 1 & 3 & 2 \\ -1 & 2 & 1 \\ 2 & 1 & 0 \end{bmatrix}$

   a. Is $A \cdot A^{-1} = I = A^{-1} \cdot A$?
   b. Is $(A^t)^{-1} = (A^{-1})^t$?

5. Given $C = \begin{bmatrix} 1 & 2 \\ -1 & 4 \end{bmatrix}$, find C^{-1}.

6. Given $A = \begin{bmatrix} 3 & 2 & 6 \\ 1 & 1 & 2 \\ 2 & 2 & 5 \end{bmatrix}$ and $B = \begin{bmatrix} 1 & 2 & -2 \\ -1 & 3 & 0 \\ 0 & -2 & 1 \end{bmatrix}$, determine whether or not $B = A^{-1}$.

10-7 LINEAR SYSTEMS IN THREE VARIABLES

Using matrices, we can find the solution set for a system of linear equations in three variables by the procedures outlined in the following example (procedures which are useful in computer programming).

EXAMPLE   Let

$$x + y - 2z = 2$$
$$2x + z = 3$$
$$x - y + z = 1$$

The coefficients on the three variables x, y, and z are the entries found in the 3 × 3 matrix below. (In the second equation, the coefficient on y is 0.)

Multiplying the entries by the variables, we get

$$\begin{bmatrix} 1 & 1 & -2 \\ 2 & 0 & 1 \\ 1 & -1 & 1 \end{bmatrix} \cdot \begin{bmatrix} x \\ y \\ z \end{bmatrix} = \begin{bmatrix} x + y - 2z \\ 2x + 0y + z \\ x - y + z \end{bmatrix}$$

Thus, the 3 × 3 matrix and the 3 × 1 matrix are both conformable for multiplication and yield the left-hand side of the three linear equations under consideration.

The 3 × 3 matrix yields a determinant not equal to 0, and therefore a solution exists for these equations. Note that the original 3 × 3 matrix is identical to the one given in Example 3, page 228.

We further note that

$$\begin{bmatrix} 1 & 1 & -2 \\ 2 & 0 & 1 \\ 1 & -1 & 1 \end{bmatrix} \cdot \begin{bmatrix} x \\ y \\ z \end{bmatrix} = \begin{bmatrix} 2 \\ 3 \\ 1 \end{bmatrix} \text{ where } \begin{bmatrix} 2 \\ 3 \\ 1 \end{bmatrix}$$

is found on the right hand side of the equal signs in the three given equations.

Recall that

$$A^{-1} = \tfrac{1}{4} \cdot \begin{bmatrix} 1 & 1 & 1 \\ -1 & 3 & -5 \\ -2 & 2 & -2 \end{bmatrix}$$

We have verified that $A^{-1} \cdot A = I$. Now if

$$\begin{bmatrix} 1 & 1 & -2 \\ 2 & 0 & 1 \\ 1 & -1 & 1 \end{bmatrix} \cdot \begin{bmatrix} x \\ y \\ z \end{bmatrix} = \begin{bmatrix} 2 \\ 3 \\ 1 \end{bmatrix}$$

we can express this in symbolic notation as $A \cdot B = C$, where $A =$ the matrix of the coefficients, $B =$ the unknown matrix and $C =$ the constant matrix. To solve for B, we can write $B = A^{-1} \cdot C$. Since $A \cdot B = C$,

$$\text{then} \quad A^{-1}(A \cdot B) = A^{-1} \cdot C$$
$$(A^{-1} \cdot A) \cdot B = A^{-1} \cdot C$$
$$I \cdot B = A^{-1} \cdot C$$
$$B = A^{-1} \cdot C$$

Expressing $B = A^{-1} \cdot C$ in terms of their corresponding matrices we have

$$\begin{bmatrix} x \\ y \\ z \end{bmatrix} = \frac{1}{4} \begin{bmatrix} 1 & 1 & 1 \\ -1 & 3 & -5 \\ -2 & 2 & -2 \end{bmatrix} \cdot \begin{bmatrix} 2 \\ 3 \\ 1 \end{bmatrix} = \frac{1}{4} \begin{bmatrix} 6 \\ 2 \\ 0 \end{bmatrix} = \begin{bmatrix} \frac{3}{2} \\ \frac{1}{2} \\ 0 \end{bmatrix}$$

and the solution set for this system of linear equations is $\{\frac{3}{2}, \frac{1}{2}, 0\}$, where $x = \frac{3}{2}$, $y = \frac{1}{2}$, and $z = 0$. Substitution of these values for x, y and z into the system of linear equations is left to the student.

Elimination is another method of finding the solution set for a system of three linear equations in three variables. The following algorithm, suitable for use on computers, yields the identity matrix and the values for x, y, and z. (Before proceeding, recall that $A^{-1} \cdot A = I$ (page 227).) Consider the system

$$3x + 2y - 4z = -5$$
$$x + y + z = 6$$
$$x - y - z = -4$$

First-Column Operations

We can write:

Pivotal
column

$$\begin{array}{c} \text{Pivotal} \rightarrow \\ \text{row} \end{array} \begin{pmatrix} 3 & 2 & -4 \\ 1 & 1 & 1 \\ 1 & -1 & -1 \end{pmatrix} \begin{pmatrix} -5 \\ 6 \\ -4 \end{pmatrix}$$

First, we need 1 at a_{11}.

$$\begin{pmatrix} 3 & 2 & -4 & \vdots & -5 \\ 1 & 1 & 1 & \vdots & 6 \\ 1 & -1 & -1 & \vdots & -4 \end{pmatrix}$$ Divide 1st row by a_{11}.

$$\begin{pmatrix} 1 & \frac{2}{3} & -\frac{4}{3} & \vdots & -\frac{5}{3} \\ 1 & 1 & 1 & \vdots & 6 \\ 1 & -1 & -1 & \vdots & -4 \end{pmatrix}$$ We now want zero at a_{21}. Multiply 1st row by $-a_{21}$, add it to 2nd row, and replace entries in the 2nd row with the answer. (We will multiply by a negative of the entry where we need a zero in the identity matrix.)

$$\begin{pmatrix} 1 & \frac{2}{3} & -\frac{4}{3} & \vdots & -\frac{5}{3} \\ 0 & \frac{1}{3} & \frac{7}{3} & \vdots & \frac{23}{3} \\ 1 & -1 & -1 & \vdots & -4 \end{pmatrix}$$

$$\begin{pmatrix} 1 & \frac{2}{3} & -\frac{4}{3} & \vdots & -\frac{5}{3} \\ 0 & \frac{1}{3} & \frac{7}{3} & \vdots & \frac{23}{3} \\ 0 & -\frac{5}{3} & \frac{1}{3} & \vdots & -\frac{7}{3} \end{pmatrix}$$ We now need zero at a_{31}. Multiply 1st row by $-a_{31}$ and add to 3rd row. Then replace 3rd row with the answer.

Second-Column Operations

Now the pivotal row is the second row, and the pivotal column is the second column.

$$\begin{pmatrix} 1 & \frac{2}{3} & -\frac{4}{3} & \vdots & -\frac{5}{3} \\ 0 & 1 & 7 & \vdots & 23 \\ 0 & -\frac{5}{3} & \frac{1}{3} & \vdots & -\frac{7}{3} \end{pmatrix}$$ To change a_{22} to 1, divide 2nd row by a_{22}.

$$\begin{pmatrix} 1 & 0 & -\frac{18}{3} & \vdots & -\frac{51}{3} \\ 0 & 1 & 7 & \vdots & 23 \\ 0 & -\frac{5}{3} & \frac{1}{3} & \vdots & -\frac{7}{3} \end{pmatrix}$$ To get zero at a_{12}, multiply the 2nd row by $-a_{12}$, add to 1st row, and replace 1st row with the answer.

$$\begin{pmatrix} 1 & 0 & -\frac{18}{3} & \vdots & -\frac{51}{3} \\ 0 & 1 & 7 & \vdots & 23 \\ 0 & 0 & 12 & \vdots & 36 \end{pmatrix}$$ To get zero at a_{32}, multiply 2nd row by $-a_{32}$, add to 3rd row, and replace 3rd row with answer.

Third-Column Operations

Now the pivotal row is third row, and the pivotal column is the third column

$$\begin{pmatrix} 1 & 0 & -6 & \vdots & -17 \\ 0 & 1 & 7 & \vdots & 23 \\ 0 & 0 & 1 & \vdots & 3 \end{pmatrix}$$ For a 1 at a_{33}, divide 3rd row by a_{33}.

$$\begin{pmatrix} 1 & 0 & 0 & \vdots & 1 \\ 0 & 1 & 7 & \vdots & 23 \\ 0 & 0 & 1 & \vdots & 3 \end{pmatrix}$$ For a zero at a_{13}, multiply 3rd row by $-a_{13}$, add to 1st row, and replace 1st row with answer.

$$\begin{pmatrix} 1 & 0 & 0 & \vdots & 1 \\ 0 & 1 & 0 & \vdots & 2 \\ 0 & 0 & 1 & \vdots & 3 \end{pmatrix}$$ Finally, we need a zero at a_{23}. Multiply the 3rd row by $-a_{23}$, add the result to the 2nd row, and replace the 2nd row with the answer.

Identity   values
Matrix   for x, y
    and z

Thus, the solution set for

$$3x + 2y - 4z = -5$$
$$x + y + z = 6$$
$$x - y - z = -4$$

is {1, 2, 3}, where $x = 1$, $y = 2$, and $z = 3$.

Note that it is sometimes possible for the element a_{11} to contain a zero coefficient of x. To get an element $a_{11} \neq 0$, it is possible to exchange the first row with a row that has an x whose coefficient is not zero. For example, given

$$\begin{pmatrix} y + 3z \\ 2x + 3y - z \\ -x + z \end{pmatrix}$$

we may exchange the 1st and 2nd rows, yielding

$$\begin{pmatrix} 2x + 3y - z \\ 0 + y + 3z \\ -x + 0 + z \end{pmatrix}$$

Exercises 10-5

1. Given:

$$x + y - z = -1$$
$$2x - y + z = 2$$
$$x + y + 2z = 0$$

a. If the above system is written in terms of matrices A, X, and B, where $AX = B$, identify A, X, and B.
b. Find A^{-1}.
c. Solve the system using A^{-1}.

2. Find the solution sets of the following. (If the solution set is the null set, identify it.) Use either method discussed in Section 10-7.

a. $y + 2z = 3$
   $3x + z = 4$
   $x + y = 2$

b. $x - y + z = 3$
   $2y - x - 2z = 2$
   $x - y + 2z = 4$

c. $y - x + 2z = 3$
   $z - x = 2$
   $x + 2y + z = 8$

3. In this chapter we have described several methods for solving simultaneous equations. Study the Basic program that follows and describe how it solves these types of equations.

    The REM, or REMarks, statement contains hints on how this program executes. The REM statement may be separated from another Basic statement by a colon (:). A REM statement is not executable, being typically used for comments about other statements in the program.

    Look at statement 130. A GOSUB instruction transfers program control to a subroutine that begins at a specified line number (in this case 1000). When the computer encounters a RETURN statement (1130 in this program), the program will return to the statement that directly follows the GOSUB.

```
10 REM PROGRAM TO SOLVE 3 SIMULTANEOUS EQUATIONS
20 PRINT "ENTER THE THREE COEFFICIENTS IN ORDER X,Y,Z"
30 PRINT "NEXT ENTER THE CONSTANT"
40 DIM A(5,5), X(5,5)
45 INPUT "FIRST EQUATION      ";A(1,1),A(1,2),A(1,3),A(1,4)
50 INPUT "SECOND EQUATION ";A(2,1),A(2,2),A(2,3),A(2,4)
60 INPUT "THIRD EQUATION    ";A(3,1),A(3,2),A(3,3),A(3,4)
70 REM LOAD ARRAY X
80 FOR I = 1 TO 4
```

```
90 FOR J = 1 TO 4
100 X(I,J)=A(I,J): REM SET UP A MEMORY MATRIX FOR REFERENCE
110 NEXT J
120 NEXT I
130 GOSUB 1000
140 REM FIND THE DETERMINANT
150 D = Q: REM RETURN VALUE OF SUBROUTINE AT LINE 1000
160 A(1,1)=A(1,4)
170 A(2,1)=A(2,4)
180 A(3,1)=A(3,4)
190 GOSUB 1000: REM SOLVING MATRIX WITH ROW 1 SUBSTITUTED
200 X=Q/D: REM X IS NUMERATOR DIVIDED BY DETERMINANT
210 A(1,2)=A(1,4)
220 A(2,2)=A(2,4)
230 A(3,2)=A(3,4)
240 GOSUB 1000
250 Y=Q/D
255 A(1,3)=A(1,4)
260 A(2,3)=A(2,4)
270 A(3,3)=A(3,4)
290 GOSUB 1000
300 Z=Q/D
310 REM OUTPUT ROUTINE
320 PRINT
330 PRINT "THE ORIGINAL EQUATIONS WERE"
340 PRINT
350 PRINT X(1,1);"X + ";X(1,2);"Y + ";X(1,3);"Z = ";A(1,4)
355 PRINT
360 PRINT X(2,1);"X + ";X(2,2);"Y + ";X(2,3);"Z = ";A(2,4)
365 PRINT
370 PRINT X(3,1);"X + ";X(3,2);"Y + ";X(3,3);"Z = ";A(3,4)
375 PRINT
377 PRINT "SOLUTION"
380 PRINT "X = ";X;"   Y = ";Y;"   Z = ";Z
390 END
1000 D1 = A(1,1)*A(2,2)*A(3,3):REM MULTIPLY
1010 D2 = A(1,2)*A(2,3)*A(3,1)
1020 D3 = A(1,3)*A(2,1)*A(3,2)
1030 D4 = A(1,3)*A(2,2)*A(3,1)
1040 D5 = A(1,2)*A(2,1)*A(3,3)
1050 D6 = A(1,1)*A(2,3)*A(3,2)
1060 Q = D1 + D2 + D3 − D4 − D5 − D6          SUBROUTINE
1070 REM RESTORE ORIGINAL MATRIX
1080 FOR I = 1 TO 4
```

```
1090 FOR J = 1 TO 4
1100 A(I,J)=X(I,J)
1110 NEXT J
1120 NEXT I
1130 RETURN
1140 REM****END OF SUBROUTINE
```

a. Why is a subroutine useful in finding the solution set to simultaneous equations?
b. In statement 40 we set up two matrices in one statement. Why do we need the second matrix?
c. When entering data and running the program, you must use commas to separate the coefficients and the constant. If you have a computer available, type in this program, run it, and verify the answers to exercises 10-5. What message did you get from the program when running Exercise 10-5, Problem 2c?

10-8 ARRAYS AND COMPUTERS

Section 10-1 illustrates a typical data-processing situation in which arrays are very useful. Most commonly used computer languages provide techniques for operations on arrays. An array is a collection of elements each of which has attributes identical to the rest of the elements in the array. If

$$A = \begin{bmatrix} 1.3 & 4.1 \\ 0.8 & -6.2 \end{bmatrix}$$

we can describe A as a 2×2 array containing decimal numbers, each of which contains one decimal fraction.

Using the subscript notation employed earlier in this chapter to identify each element in a matrix, we can write for array A:

$$a_{11} = 1.3 \quad a_{12} = 4.1$$
$$a_{21} = 0.8 \quad a_{22} = -6.2$$

To access any of the elements in this array, it is a simple matter to write into the computer program the array name followed by the appropriate subscripts enclosed in parentheses. Thus, $A(1, 2)$ is a reference to the element a_{12}, or the number 4.1.

If a matrix appeared as

$$[1, 2, 3, 4]$$

or

$$\begin{bmatrix} 1 \\ 2 \\ 3 \\ 4 \end{bmatrix}$$

it could be expressed as a one-dimensional array containing integers.

10-8.1 Two- and Three-Dimensional Arrays

In an array declared ARRAY (2, 4) the data items would be collected and coded for the computer in the following manner:

COLUMNS

ROWS   ARRAY (1, 1)   ARRAY (1, 2)   ARRAY (1, 3)   ARRAY (1, 4)
          ARRAY (2, 1)   ARRAY (2, 2)   ARRAY (2, 3)   ARRAY (2, 4)

The first subscript refers to rows, and the second subscript refers to columns.

We can think of ARRAY (1, 2) as an element which resides in the first row, second column. If the following values were assigned,

1 ② 3 4
5 6 7 8

then ARRAY subscripted by (1, 2) would be a reference to the 2 circled above.

Another way of expressing a two-dimensional array is through the use of a "tree" diagram in Fig. 10-1, which could be declared TABLE (3, 4).

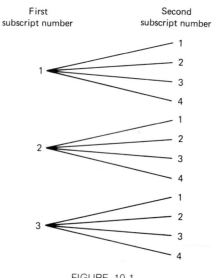

FIGURE 10-1

For TABLE (3, 4), we would have as elements:

TABLE (1, 1)   TABLE (1, 2)   TABLE (1, 3)   TABLE (1, 4)
TABLE (2, 1)   TABLE (2, 2)   TABLE (2, 3)   TABLE (2, 4)
TABLE (3, 1)   TABLE (3, 2)   TABLE (3, 3)   TABLE (3, 4)

A three dimensional array requires three subscripts. For example, AVE (2, 2, 3) can be considered as an array containing two groups separated into two lists with each list containing three data items.

AVE (1, 1, 1)   AVE (1, 1, 2)   AVE (1, 1, 3)
AVE (1, 2, 1)   AVE (1, 2, 2)   AVE (1, 2, 3)
AVE (2, 1, 1)   AVE (2, 1, 2)   AVE (2, 1, 3)
AVE (2, 2, 1)   AVE (2, 2, 2)   AVE (2, 2, 3)

This is expressed in tree diagram form in Fig. 10-2.

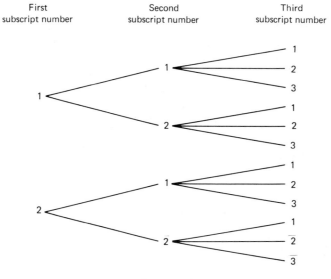

FIGURE 10-2

Using the array AVE (2, 2, 3), we can present a practical application of Fig. 10-2 in average test scores for students in the last three years of two secondary schools:

	School A Average			School B Average	
Year	*Midterm*	*Final*	*Year*	*Midterm*	*Final*
10	80.0	80.4	10	83.1	83.2
11	81.2	81.3	11	82.0	82.5
12	82.0	82.1	12	82.6	83.0

Arranging this data in another manner, we have

	School A Average				School B Average		
Year	10	11	12	Year	10	11	12
Midterm	80.0	81.2	82.0	Midterm	83.1	82.0	82.6
Final	80.4	81.3	82.1	Final	83.2	82.5	83.0

Note that the School A, School B, years 10, 11, 12, midterm, and final are not elements in either array. The only elements in the arrays are average grades. These are given as homogeneous elements—in this case, numbers containing decimal fractions.

The eleventh year final average for School A is referenced as AVE (1, 2, 2), or first group, second row, second column. How is the twelfth year midterm average for School B referenced in AVE?

10-8.2 Manipulating Data in Arrays

In Fortran the DIMENSION statement is used to reserve an area in the computer's storage for arrays of the size specified by the information found in the statement. Consider, for example,

DIMENSION A(5), B(2, 5), C(2, 3, 4).

When the statement is executed the computer will reserve 5 storage locations to array A, 10 (or 2 × 5) storage locations to array B, and 24 (or 2 × 3 × 4) storage locations to array C. The equivalent statements in PL/I are:

DECLARE A (5);
DECLARE B (2, 5);
DECLARE C (2, 3, 4);

where array A will contain 5 storage locations, B 10 storage locations, and C 24 storage locations.

Before manipulating data in matrices and arrays in a computer program, we must briefly examine the DO statement. The DO statement is one of the most powerful in the Fortran and PL/I languages. It makes it possible to repeat an instruction or a series of instructions by means of relatively simple, easily readable code. A simple form of the Fortran DO statement appears as

$$DO\ n\ i = n_2, n_3$$

where

n is the statement number of the last statement in the loop
i is the index variable (usually a single letter I, J, K, L, M, or N)
n_2 is the first value of the index variable
n_3 is the last value of the index variable

For example, in

$$N = 0$$
$$DO\ 10\ I = 1,\ 5$$
$$10\ N = N + 1$$
$$\equiv \} \text{ next statements in program}$$

N = 0 means that N is initialized at 0; N = N + 1 means that N is re-placed by itself + 1; the number 10 indicates statement number 10; and I is the index variable, which is set at 1 for the first execution of statement 10. Note that statement 10 will be executed 5 times. Each time statement 10 is executed, I will increase in value by 1 until it reaches 5. After I = 5, the program will continue on to the next statements in the program.

Next, consider

$$\text{DIMENSION A(10), B(10), C(10)}$$
$$DO\ 15\ I = 1,\ 10$$
$$15\ C(I) = A(I) + B(I)$$

The number 15 is the number of a Fortran statement that is to be exe-cuted 10 times.

In

$$PI = 3.14$$
$$R = 1.0$$
$$DO\ 25\ J = 1,\ 20$$
$$AREA = PI^* R^{**}\ 2$$
$$25\ R = R + 1.0$$

the DO statement causes two statements

$$AREA = PI^* R^{**}\ 2 \text{ and}$$
$$25\ R = R + 1.0$$

to be executed 20 times.

Any number of statements may appear between the DO statement and the statement number that is referenced by it.

Exercises 10-6

1. The following Fortran code contains a DO loop and an IF statement. Explain the activities of this coded routine.

$$HIGH = X(1)$$
$$10\ DO\ 40\ K = 2,\ 50$$
$$20\ IF\ (X(K) - HIGH)\ 40,\ 40,\ 30$$
$$30\ HIGH = X(K)$$
$$40\ CONTINUE$$

(The CONTINUE statement is a "dummy" statement, providing a common finishing point in a DO loop. The computer will return to state-ment 10 and execute all statements 10, 20, 30, or 40 while K = 2 and

$K \leq 50$. Note that K is initialized at 2 instead of 1 and X is an array with 50 elements.)

2. *Matrix Addition Problem.* Given three one-dimensional arrays (for example, IARAY1, IARAY2, IANS) each of which contains 10 elements, do the following, using Fortran:

   a. Initialize the values into the first two arrays so that the first elements = 1, the second elements = 2, etc.

   b. Sum up corresponding elements of the first two arrays, and place the result in the corresponding position of the third array.
   Hint: Study the subscripting of array X in problem 1.

3. *Reverse Elements in a Matrix.* Given a one dimension array (IARRAY) consisting of 10 elements, do the following:

   a. Initialize IARRAY so that IARRAY(1) = 1, IARRAY(2) = 2, etc.

   b. Invert IARRAY; that is, reverse IARRAY(1) and IARRAY (10), IARRAY(2) and IARRAY(9), etc.

4. Using DO loops, write a Fortran routine to find the solution set for

$$3x + 2y - 4z = -5$$
$$x + y + z = 6$$
$$x - y - z = -4$$

   In writing the DO loops, use the algorithm for elimination discussed on pages 231–232.

5. The Basic language uses FOR NEXT loops in a similar manner to the Fortran DO loop. Study the following Basic program and describe its function. (ARAY is a single-dimensional array containing integers.)

```
200 FOR J = 1 TO 160
210 FOR H = 1 TO 159
220 IF ARAY(H) > ARAY(H+1) THEN GO TO 230 ELSE GO TO 260
230 TEMP = ARAY(H)
240 ARAY(H) = ARAY(H+1)
250 ARAY(H+1) = TEMP
260 NEXT H
270 NEXT J
```

   How many elements are in ARAY?

6. The Cobol programming language uses its OCCURS clause to create arrays. Following is a bank depositor's record containing an array of 20 elements, each representing a possible deposit or withdrawal made within a given month.

```
01 DEPOSITOR-RECORD.
    02 NAME                          PICTURE X(19).
    02 SOCIAL-SECURITY-NUMBER        PICTURE X(11).
    02 ADDRESS                       PICTURE X(30).
```

02 TRANSACTION OCCURS 20 TIMES.
    03 DEPOSIT   PICTURE 9999V99.
    03 WITHDRAWAL   PICTURE 9999V99.

What is accomplished by the programmer in the following routine?

```
              MOVE 20 TO N.
LOOP.         MOVE ZEROS TO DEPOSIT(N).
              SUBTRACT 1 FROM N.
              IF N = ZERO GO TO NEXT-LINE.
              GO TO LOOP.

    NEXT LINE.
    _____
    _____
    _____          other statements in program
    _____
    _____
```

7. Write the Fortran coding to take the overall average for schools A and B for grades 10, 11, 12, including midterm and final averages. (See page 238.

11

LINEAR
PROGRAMMING

11-1 THE CHARACTERISTICS
OF LINEAR PROGRAMMING

Linear programming is concerned with nonnegative solutions to systems of linear equations. It is sometimes referred to as the algebra of nonnegative numbers and involves the use of various methods or algorithms to solve problems. In this chapter we will use algebraic, matrix, and graphing methods as problem-solving tools for linear programming problems.

Linear programming problems have the following characteristics:

a. A desired objective, such as maximum profit, minimum cost, or minimum time.

b. A large number of variables to be manipulated at the same time, such as ingredients in a gasoline mixture, money, plant space, products, man- or machine-hours.

c. Constraints or restrictions upon variables, such as the number of man-hours allowable by union contract, output possible from a machine, the demand for a product in an open market. (Without restrictions on the variables, most problems are trivial.)

d. Interaction among the variables; for example, in determining which products to manufacture, the maximum profits, man-hours, plant capacity, and costs for all products must be considered. Plants have limited capacities, some types of skilled labor are more costly than others, certain kinds of raw materials are more difficult to acquire than others, and so on. For reasons such as these, products *compete* for resources, and linear programming methods can be used to determine the most profitable approach to production.

The word "linear" in linear programming implies that the variables are linear, i.e., they can be expressed in the relationships found in linear equations. Some typical problems in linear programs are:

1. *Mixture Problems.* (Example: Food prices vary from week to week and a housewife must purchase foods that meet nutritional requirements as well as financial limitations.)
2. *Production Problems.* (Example: Scheduling machines to get the greatest output for the least cost.)
3. *Transportation Problems.* (Example: Products are stored in several warehouses, from which they must be distributed to various retail stores and individual customers in the minimum amount of time and in the least expensive manner.)

Linear programming is an important tool in finding the optimum values of linear functions subject to specified constraints.

11-2 MAXIMIZING PROFIT

Using nonnegative values, consider a manufacturer of fuel-injection systems making two different types of systems: System *A* for power boats and System *B* for automobiles. There are two separate plants, Plant 1 and Plant 2, each capable of producing both types of fuel-injection systems. After the parts are produced, they are packaged and shipped to another plant for product-test and assembly. Assume that

1. System *A* requires 60 hours in Plant 1 or 70 hours in Plant 2.
2. System *B* requires 80 hours in Plant 1 or 40 hours in Plant 2.
3. Under union rules, for a labor force of constant size, only 480 hours of work can be performed in plant 1 and 280 hours of work in Plant 2 each week.
4. For present market conditions, it is possible to realize a profit of 30 dollars on System *A* and 20 dollars on System *B*.

Fig. 11-1 shows the relationship of hours to profits for each of the two systems in each of the two plants.

	System A	System B	Union hours
Plant 1 hours	60	80	480
Plant 2 hours	70	40	280
Profits	30	20	

FIGURE 11-1

Let

A = the number of marine fuel-injection systems produced

B = the number of automobile fuel-injection systems produced

P = total profit

Because the number of fuel-injection systems produced is a nonnegative number, $A \geq 0$ and $B \geq 0$. Then for plant 1

$$60A + 80B \leq 480$$

where System A requires $60A$ hours of labor, System B requires $80B$ hours of labor, and 480 represents a limitation in terms of hours imposed by union rules. For Plant 2, this works out as

$$70A + 40B \leq 280$$

For both plants, total profit can be expressed as

$$30A + 20B = P$$

To realize maximum profit, how many of each of the fuel-injection systems should each plant manufacture?

Because we are interested only in nonnegative numbers, we can examine these considerations in the first quadrant of a graph. Let the x-axis represent System A and the y-axis System B. To plot the line for Plant 1, set the number of systems produced equal to 480, or

$$60A + 80B = 480$$

Next, set $A = 0$ and $B = 6$; then set $B = 0$ and $A = 8$. Next connect the two points (0, 6) and (8, 0). Thus,

If System $A = 0$, then System $B = 6$.

If System $B = 0$, then System $A = 8$.

The shaded area in Fig. 11-2 contains all the points that satisfy the inequality $60A + 80B \leq 480$ and describes the constraints or restrictions on the production in Plant 1 that are also given in the inequality.

In a similar manner, we can plot a graph for Plant 2.

The shaded area in Fig. 11-3 contains all the points that satisfy the inequality $70A + 40B \leq 280$ and describes the constraints for Plant 2.

In Fig. 11-4 we have combined Fig. 11-2 with Fig. 11-3.

The heavily shaded portion now contains all of the solutions common to

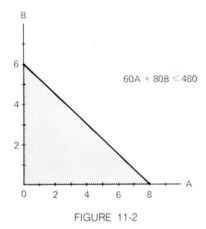

FIGURE 11-2

both inequalities. This area will be called the feasible region and represents the combination of fuel-injection systems produced by both plants under the given constraints.

Our objective in graphing these inequalities is to find a combination which provides the greatest profit (or the highest value for P) in

$$30A + 20B = P$$

In other words, we will try to maximize profits under the two constraints given.

Expressing our profit equation in terms of the formula for a straight line $y = mx + b$, we can substitute B for y and write

$$B = -\tfrac{3}{2}A + \tfrac{P}{20}$$

The slope $-\tfrac{3}{2}$ is a constant that can be moved across the feasible region. The lines S_1, S_2, and S_3 (Fig. 11-5) are parallel and all have the slope $-\tfrac{3}{2}$.

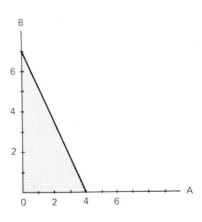

FIGURE 11-3

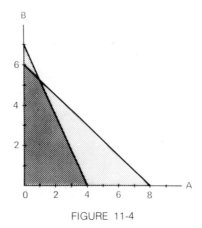

FIGURE 11-4

S_1 lies entirely within the feasible region. However, there are many points outside of S_1 that also lie within the feasible region and represent a greater value for P than those points on S_1. S_3 contains no points in the feasible region and therefore represents no realistic value for P. All points on the line S_3 are beyond the production capability of both plants.

As the line with the slope $-\frac{3}{2}$ moves out from the origin, the maximum profit will be reached when that line touches the last feasible point in the solution set. S_2 represents the maximum feasible profit line and contains a point that can be found by solving the two simultaneous equations

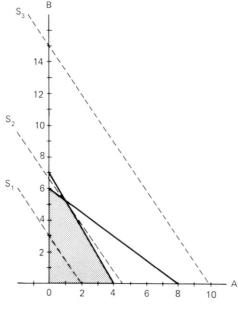

FIGURE 11-5

$$60A + 80B = 480$$
$$70A + 40B = 280$$

Solving, we obtain $A = 1$ and $B = 5.25$. This answers the question, "How many of each of the fuel-injection systems per week should each plant manufacture?" (Consider that one-fourth of the work will be done towards a complete fuel-injection System B near the end of each week and that every four weeks we will have twenty-one automobile fuel-injection systems.)

Substituting 1 for A and 5.25 for B in

$$30A + 20B = P$$

we get

$$30.1 + 20 \cdot 5.25 = P$$

Thus, $P = \$135.00$, which is the maximum profit under the given constraints.

Observe that the line with slope $-\frac{3}{2}$ will also touch points (0, 6) and (4, 0). As an exercise, determine whether there would be a profit in dropping production for either System A and System B.

Exercises 11-1

1. Using Fig. 11-5, locate the following points and determine whether or not they are in the feasible region. (Let the first number represent a value for System A, the second number represent a value for System B.)

   a. (1, 2)   b. (5, 5)   c. (3, 4)   d. (2, 3)   e. (6, 4)

2. If the number of unused and overtime hours for each plant is as given below, complete the tables for A and B.

   a.

	Unused Hours	A	Units	B
Plant 1	40	2		
Plant 2	30			1

   b.

	Over Time Hours	A	Units	B
Plant 1	100			5
Plant 2	80	4		

FIGURE 11-6

Hint:
For Plant 1 $A + B = 480 -$ unused hours
For Plant 2 $A + B = 280 -$ unused hours

3. Using the same intersection of the simultaneous equations for Fig. 11-2 and Fig. 11-3, find P (the profit) if
   a. $10A + 20B = P$
   b. $15A + 10B = P$
   c. $70A + 80B = P$

4. Graph the following, shade the feasible region, draw the maximum profit line, and solve the simultaneous equations to maximize. (Find the maximum value for P.)
   a. $7x + 3y = P$
      where the constraints are
      $2x + 4y \leq 80$
      $5x + 3y \leq 150$
   b. $3x + 4y = P$
      where the constraints are
      $x + y \leq 120$
      $x + 2y \leq 160$

5. Two induction motors are made in three plants. The time requirements for each plant are shown in Fig. 11-7.

	Plant*			
	1	2	3	Profit Per Unit
Motor A	4	5	7	$18
Motor B	8	5	4	$12
Max. hrs. available	320	250	280	—

* Hours to make a unit.

FIGURE 11-7

Graph the three equations and, using $P = 18A + 12B$, indicate the best possible combination of manufacturing these two motors that will yield the maximum profit. (A and B are units to be manufactured.)

6. Assume the profit on item A is $5 and the profit on item B is $7 and a maximum production is 15 for item A and 9 for item B; using matrix algebra, we can then find the maximum profit as follows:

$$[15\ 9] \begin{bmatrix} 5 \\ 7 \end{bmatrix} = 75 + 63 = \$138$$

   a. Using matrix multiplication, show the profit involved in the manufacture of the fuel-injection systems described on page 243.
   b. In the same problem recall that
      $$60A + 80B = 480$$
      $$70A + 40B = 280$$

and complete

$$[1 \quad 5.25] \begin{bmatrix} \\ \end{bmatrix} = 480$$

$$[1 \quad 5.25] \begin{bmatrix} \\ \end{bmatrix} = 280$$

c. Write the appropriate Fortran statement to perform the multiplication indicated in part b).

7. Refer to Fig. 11-5 for the following problems.

a. If the solution to a linear programming problem must occur at an intersection, and if S_2 in Fig. 11-5 had the slope $-\frac{3}{4}$, where would the maximum profit be found?

b. If the slope S_2 were almost vertical (very large) and were moved on the Cartesian plane from right to left, which point would it touch first?

11-3 MINIMIZING COSTS

To reduce pollution in a nearby river, factory management has decided to treat its waste with two standard products. Each of these standard products contains three important antipollutants in varying amounts. The respective amounts per gallon of the three antipollutants in Product A are 10-9-2 and in Product B, 5-15-9. Minimum hourly requirements for each antipollutant are respectively 30-45-18. Product A costs 80 cents per gallon and Product B costs 60 cents per gallon. The management wishes to find the most inexpensive mixture of A and B that meets the minimum requirements.

Figure 11-8 displays the antipollution problem, showing the amounts of antipollution items for each product, the hourly requirements of each of the three items, and the cost per gallon. Again, we are interested only in nonnegative numbers. Therefore, $A \geq 0$ and $B \geq 0$.

While it would be desirable to get a mixture of A and B that contains the exact hourly requirement, the management is willing to accept an

Antipollutants by units per gallon

Product	Item 1	Item 2	Item 3	Cost
A	10	9	2	0.80
B	5	15	9	0.60
Required/hour	30	45	18	

FIGURE 11-8

excess, rather than a deficiency, of one or two antipollutants. Therefore, we can write

1. $10A + 5B \geq 30$
2. $9A + 15B \geq 45$
3. $2A + 9B \geq 18$

Figure 11-9 expresses the problem in graphical form.
In the figure, the line

1. connecting points (3, 0) and (0, 6) represents the equation $10A + 5B = 30$ (item 1)
2. connecting points (5, 0) and (0, 3) represents the equation $9A + 15B = 45$ (item 2)
3. connecting points (9, 0) and (0, 2) represents the equation $2A + 9B = 18$ (item 3)

The line drawn for item 1 indicates that three gallons of Product *A* or six gallons of Product *B* provide the exact minimum for item 1. But three gallons of Product *A* yield 27 units of item 2, which represent a deficiency of 18 units for the hourly requirement, and six gallons of Product B yield 90 units of item 2, which represent a 45-unit excess for the hourly requirement.

An analysis of the other two lines yields similar results.

The shaded area represents a feasible region which contains points that represent combinations of the three items yielding, at the very least, the minimum hourly requirements. The shaded area contains all those points which represent solutions to all three inequalities. The feasible region is bounded in part by the lines *JK, KL,* and *LM* in Fig. 11-10.

Our objective in graphing these three lines is to find a combination which provides the least possible cost, or lowest value, for *C* where $.80A + .60B = C$. Stated in other terms, minimize $.80A + .60B = C$

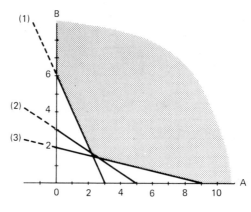

FIGURE 11-9

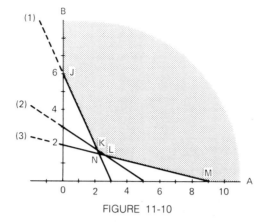

FIGURE 11-10

under the following constraints:

$$10A + 5B \geq 30$$
$$9A + 15B \geq 45$$
$$2A + 9B \geq 18$$

Expressing our cost equation in terms of the formula for a straight line (and substituting B for y in $y = mx + b$), we have

$$B = -\frac{4}{3}A + \frac{C}{.60} \quad \text{(where } C \text{ represents the cost)}$$

Drawing three lines with a slope of $-\frac{4}{3}$, we note that the minimum cost should be that point which is as close as possible to the origin, but still intersects the feasible region. This eliminates consideration of line S_3 in Fig. 11-11. In other words, as the line with a slope of $-\frac{4}{3}$ moves in towards the origin, the minimum will be reached when that line touches the last feasible point in the solution set.

In Fig. 11-11, the point P on line S_1 represents 4 gallons of B and 3 gallons of A. Expressing the items in Products A and B as two 1×3

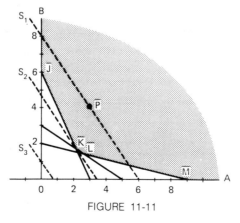

FIGURE 11-11

matrices where column 1 contains the entry for item 1, column 2 the entry for item 2 and so on, we have

$$A = [10 \quad 9 \quad 2]$$
$$B = [\ 5 \quad 15 \quad 9]$$

Next, multiply each by the appropriate number of gallons:

$$3 \times [10 \quad 9 \quad 2] = [30 \quad 27 \quad 6] \quad \text{(3 gallons for } A \text{ at point } P)$$
$$4 \times [\ 5 \quad 15 \quad 9] = [20 \quad 60 \quad 36] \quad \text{(4 gallons for } B \text{ at point } P)$$

Inspection reveals 30 + 20 = 50, or an excess of 20, units for item 1; further, there is an excess of 42 units for item 2 and 24 units for item 3.

Since we are trying to minimize cost at the same time that we are trying to get the best possible mixture, we can move the slope $-\frac{4}{3}$ closer to the origin. Inspection reveals that point K seems to be the closest point to the origin that still lies within the feasible region. To find this point, we will solve the simultaneous equations obtained by the two intersecting straight lines

$$10A + \ 5B = 30$$
$$9A + 15B = 45$$

Solving yields

$$A = \tfrac{15}{7} \text{ and } B = \tfrac{12}{7}$$

Replacing A and B with $\tfrac{15}{7}$ and $\tfrac{12}{7}$, respectively, we compute

$$10A + \ 5B \tag{1}$$
$$9A + 15B \tag{2}$$
$$2A + \ 9B \tag{3}$$

For item 1: $\dfrac{15 \cdot 10}{7} + \dfrac{12 \cdot 5}{7} = 30$ gallons (exact requirement)

For item 2: $\dfrac{15 \cdot 9}{7} + \dfrac{12 \cdot 15}{7} = 45$ gallons (exact requirement)

For item 3: $\dfrac{15 \cdot 2}{7} + \dfrac{12 \cdot 9}{7} = 19.7$ gallons, or 1.7 gallons in excess of requirements

Referring to Fig. 11-8, we note that the exact requirements are met for items 1 and 2 and that we have a small excess of 1.7 gallons for item 3 in the least expensive mixture of antipollutants at point K on S_2. As stated previously, our objective in graphing these three inequalities is to find a combination which provides the least cost for

$$C = .80A + .60B$$

in other words, to minimize C under the three constraints given. Hence, substituting $\tfrac{15}{7}$ for A and $\tfrac{12}{7}$ for B, we find that

$$C = .80(\tfrac{15}{7}) + .60(\tfrac{12}{7})$$
$$C = 2.74$$

Exercises 11-2

Refer to Fig. 11-11 for problems 1 through 4.

1. a. Find the appropriate solution set at L.
   b. Use the values for A and B at L, and compute
      1. 10A + 5B
      2. 9A + 15B
      3. 2A + 9B

2. Using values for A and B at L,

   a. Refer to Fig. 11-8 and list any excess or deficiency.
   b. Find the cost of products A and B at L using matrix algebra.

3. At what point on the graph are the requirements best met for

   a. Items 1 and 3
   b. Items 2 and 3

4. Assume that there are different vendors for products A and B and that the costs for these products vary with each vendor. Let A remain equal to $\frac{15}{7}$ and B remain equal to $\frac{12}{7}$.

   a. Find the costs of products A and B at J.
   b. Use matrix algebra and compute costs for
      1. COST = .49A + .56B
      2. COST = .50A + .70B

Note: For all linear programming problems, it is always best to first graph the constraints, find the slope, and shade the feasible region, and then

   a. For maximizing, move the slope out from the origin and determine the last feasible point in the solution set.
   b. For minimizing, move the slope in the feasible region in towards the origin and determine the last feasible point in the solution set. These points may lie at the intersection of two lines or on the x or y axis.

5. In problems 5a and 5c first draw the graphs as described above, and then minimize. (Find the minimum value for C.)

   a. .30x + .20y = C
      where the constraints are
      1. 6x + 5y ≥ 30
      2. 10x + 14y ≥ 70
      3. 6x + 20y ≥ 60

   b. How many points must be examined in order to minimize C in problem a?

c. $x + y =$ COST

where the constraints are

1. $3x + 2y \geq 30$
2. $x + 3y \geq 36$
3. $3x + 5y \geq 60$

d. If COST $= 5x + 2y$ in problem 5c, at what point on the graph is the minimum cost found?

e. How many points of intersection are in the solution set for problem 4c?

f. Design your own problem to find P at *its maximum* on one of the axes of the Cartesian plane. Approach the solution to this problem by first drawing the graph.

Use matrix methods wherever possible in the following problems:

6. Accommodations must be provided for at least 408 persons in a suburb. Buildings must be of two types, X and Y, subject to the following conditions:

   House X uses 150 units of wood, 50 of concrete, has a capacity of 12 persons, takes 400 hours to build, and costs $50,000.

   House Y uses 50 units of wood, 250 units of concrete, has a capacity of 24 persons, takes 500 hours to build, and costs $120,000.

   At least 1100 units of wood and 2750 units of concrete must be used. How many houses of each type must be built, where

   a) the cost of construction is a minimum.

   b) the time of construction is a minimum.

7. An oil company owns two different wells that produce a given kind of oil. The wells are located in different parts of the country and have different production capacities. After pumping. the oil is graded into three classes: high grade. medium. and low. The oil company has a contract to provide a plant ABC with 12 barrels of high-grade. 6 barrels of medium, and 24 barrels of low-grade oil per week. It costs the company $160 per day to run the first well and $200 per day to run the second. However, in a day's operation the first well produces 6 barrels of high-grade. 2 barrels of medium-grade and 4 barrels of low-grade. while the second well produces 2 barrels of high and medium-grade each and 12 barrels of low-grade per day. How many days a week should each well be operated in order to fulfill the company's orders most economically?

8. Write the Fortran statements to multiply the elements in a one dimensional array which contains three integer elements by a constant of 3.

12

STATISTICAL MEASUREMENT

Frequently, programmers are asked to write computer programs that will provide management with statistical information. This information generally falls into two categories:

1. Measures of central tendency
2. Measures of dispersion

Many of the formulas that provide this information can be quickly coded into computer languages such as Fortran, *PL/I*, Cobol, and Basic. These formulas are given here as problem-solving tools only; their proofs lie outside of the scope of this text.

12-1 ORGANIZING THE DATA

The collection of numerical information can lead to massive quantities of data items. If these numerical data items are to be understood easily, they must be displayed and summarized effectively. There are three basic methods of doing this:

1. The report method, which summarizes data
2. The tabular method, which arranges data items in columns and rows
3. The pictorial method, which arranges quantitative information in graphical form

12-1.1 The Report Method

The report method is a typical data processing method to present information in summary form. The report shown in Fig. 12-1 summarizes various costs and incomes.

(In millions except per share data)	Second Quarter Ended June 30		Six Months Ended June 30	
	1982*	1981	1982*	1981
Sales	$5,100.1	$3,792.0	$10,102.8	$7,191.4
Cost of sales	(3,812.7)	(3,020.2)	(7,521.5)	(5,678.1)
Other operating costs	(1,200.4)	(528.8)	(2,253.8)	(1,071.8)
Operating income	87.0	243.0	327.5	441.5
Income from affiliates — equity method	12.4	15.5	38.5	32.3
Other income	161.8	11.0	226.5	151.5
Total income from all operations	261.2	269.5	592.5	625.3
Interest and other financial income	142.4	86.0	215.7	199.5
Interest and other financial costs	(264.3)	(66.9)	(474.8)	(126.3)
Total income before taxes on income and minority interest	139.3	288.6	333.4	698.5
Provision for estimated income taxes	135.0	121.0	220.0	260.0
Total income before minority interest	4.3	167.6	113.4	438.5
Minority interest — Marathon Oil Company	—	—	29.2	—
Net income	$ 4.3	$ 167.6	$ ⌐84.2	$ 438.5
Net income per share				
Primary	$ .05	$1.89	$ .92	$4.95
Fully diluted	$ .05**	$1.80	$ .92	$4.71
Dividends paid per share	$ .50	$ .50	$1.00	$1.00

* Includes effect of Marathon Oil Company acquisition.
** Conversion of convertible debentures excluded from fully diluted computation because of anti-dilutive effects.

FIGURE 12-1 (Courtesy of United States Steel Corporation.)

For example, the category, Other Operating Costs, is presented for two time periods, summarizing costs for products, services, etc. Note that costs for communications, raw materials, and taxes are not itemized. Instead, they are grouped under this one category and displayed as total amounts in two periods of time.

12-1.2 The Tabular Method

ABC College Enrollment

Year	Enrollment
1966	1850
1967	1950
1968	2500
1969	2800
1970	3000

FIGURE 12-2

This arrangement of data items in rows and columns makes it easy to note certain characteristics of the data under consideration. In the example given above we can determine that the student enrollment at College ABC has steadily increased from 1966 to 1970. The largest increase took place in the interval of time from January 1, 1968 to December 31, 1968. Since 1968, the enrollment has increased, but at a slower rate for each successive period.

Definition 12-1

An *interval* is a measured distance between real numbers. The outside limits of an interval are real numbers greater than, less than, or excluding other real numbers.

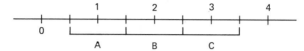

On the number line marked off in units of 1, 2, 3, and 4, the intervals marked *A*, *B*, and *C* are intervals of .5–1.5, 1.5–2.5, and 2.5–3.5, respectively. The interval *A*, for example, contains the set of real numbers from .5 to 1.5 (real numbers $\geq$ .5 and $\leq$ 1.5).

Definition 12-2

A *class* is a group of individual measurements which can be found within a given interval.

Definition 12-3

The data items or individual measurements are called *variates*.

Definition 12-4

The number of variates falling within a given class interval is called the *frequency,* or class frequency.

The pictorial method illustrates this concept.

12-1.3   The Pictorial Method

Using the enrollment figures given earlier for College *ABC*, we can create a vertical bar graph, having no space between bars. This type of vertical bar graph (Fig. 12-3) is called a *histogram,* with intervals or class boundaries for each period as points on the horizontal axis and frequencies on the vertical axis. Enrollment, the range, is expressed as a function of time, the domain.

By marking a midpoint at the top of each bar and connecting these midpoints, we can construct the *frequency polygon* shown in Fig. 12-4. The line graph in Fig. 12-4 that forms the frequency polygon is placed on top of the histogram of Fig. 12-3.

We can also separate the bars, rotate our histogram 90° clockwise, and present the same information in a horizontal bar graph. The bar graph dramatizes enrollment growth and provides a summary of the numerical information that is easily understood. Fig. 12-5 no longer ex-

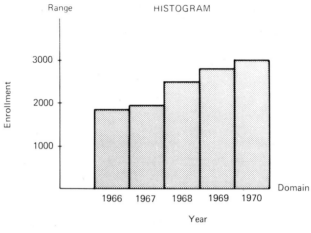

FIGURE 12-3

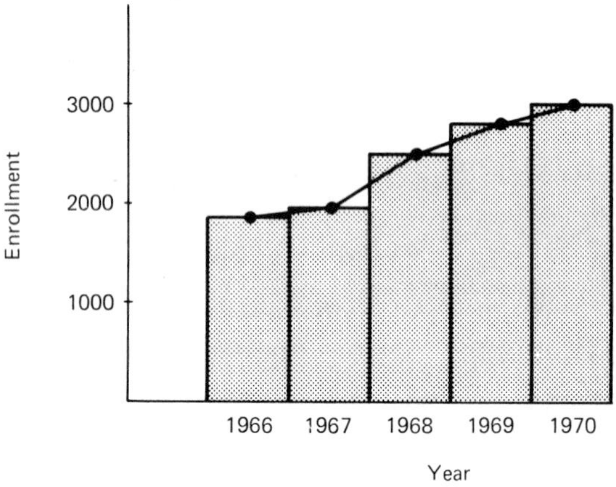

FIGURE 12-4

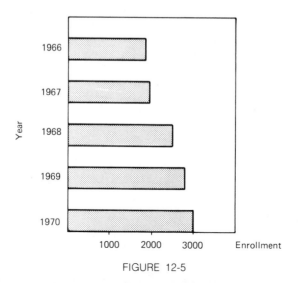

FIGURE 12-5

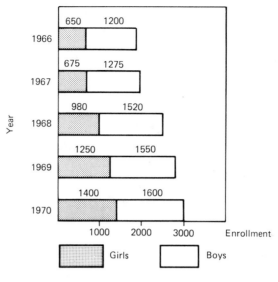

FIGURE 12-6

presses a function; rather, it graphically illustrates the dispersion of the data.

Fig. 12-6 shows a relationship of boys to girls in a component bar graph (a bar graph divided into two parts).

If we wished to make a closer comparison between the male and female enrollment figures, we could group the component bars as in Fig. 12-7. The grouped bar graph indicates that the increase in enrollment is greater for girls than for boys.

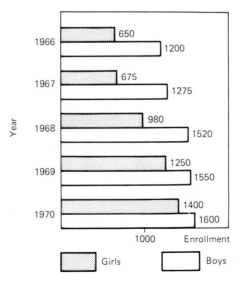

FIGURE 12-7

Fig. 12-8 is a simple display showing the universe of all students in a district. This type of graphic presentation is called a pie graph, another pictorial method. Fig. 12-8 illustrates that half of the students are elementary school students.

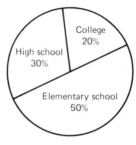

FIGURE 12-8

The selection and construction of a pictorial summary depends upon the manner in which data items are to be displayed, measured, or compared. Such data items as the number of students in a city, the number of persons over sixty-five years of age, and the amount of dollars paid for school taxes, are measured and expressed in integers and are called *discrete variables.*

A data item or variable that assumes any real number value within certain limits is called a *continuous variable.* Such data items as temperatures, average rainfalls, airplane speeds, and data items dealing with weight, time, and length are called continuous variables. These data items reflect continuous variations or changes.

Continuous data items are usually expressed in the pictorial form of a smooth curve. Fig. 12-9 displays a continuous variable.

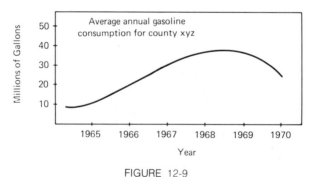

FIGURE 12-9

Exercises 12-1

1. a. Give an example of a discrete variable.
   b. Give an example of a continuous variable.

2. Of all transactions at the *XYZ* Hotel chain, 20 percent were cash, 60 percent were by one of the major credit cards, and the rest were by personal check. Draw a pie graph to show these different types of transactions.

3. The number of persons insured in a ten-year period by the *ABC* Insurance Company was:

Year	Life Insurance	Property Insurance
1961	17,322	12,476
1962	25,593	15,935
1963	46,091	30,191
1964	48,377	38,465
1965	50,633	43,870
1966	71,299	51,663
1967	85,631	72,404
1968	91,774	83,049
1969	104,392	91,223
1970	107,421	99,567

From this data:

a. Construct a component bar graph.
b. Construct a grouped bar graph.

4. The average daily temperatures at *XYZ* Airport are given as follows:

Temperatures		Number of days in year
Low	High	for temperature
20°	30°	5
30	40	35
40	50	30
50	60	65
60	70	80
70	80	70
80	90	55
90	100	25

For example, the average daily temperature ranged from 20° to 30° for five days and from 30° to 40° for 35 days.
Construct:

a. A histogram
b. A frequency polygon over the histogram.

5. The *ABC* Products Company has two different products in three areas. The products, known as *X* and *Y*, both have the same potential sales demand, but each requires a different product knowledge. Product *X* is the easiest for a salesman to understand and to explain to a customer. The estimated profit is $10 for *X* and $7 for *Y*.

Each of the three market areas has three salesmen, one of whom functions as a regional sales manager who:

1. Coordinates the activities of the other two.
2. Coordinates paper work.
3. Trains new salesmen.
4. Makes calls on customers as time allows.

Salesman C has been with the Company for only four months. The remaining eight salesmen have over two years experience. Given:

Sales Report

Region	Salesman	Product	Units
I	A Manager	X	100
		Y	95
	B	X	190
		Y	170
	C	X	100
		Y	75
II	D Manager	X	225
		Y	75
	E	X	200
		Y	100
	F	X	175
		Y	125
III	G Manager	X	115
		Y	105
	H	X	215
		Y	210
	I	X	200
		Y	190

Prepare a more meaningful report reflecting

1. Total units by product by region.
2. Manager D's low sales rate for product Y.
3. Manager G's more balanced output.
4. Profit in dollars.

12-2 SUMMATION NOTATION

With large amounts of data it is necessary to simplify the notation for grouping and summation. If a universe of terms is under consideration, such as all the students' grades in one class, we may designate the universe as X and the individual student grades as X_1 (read as "X sub one"), X_2, X_3, and so on. Then X_N denotes the last term in the universe when that universe has N terms, N being a positive integer. X_i is generally referred to as the ith term.

The symbol Σ is the *summation symbol* and indicates a sum of N terms. Thus,

$$\sum_{i=1}^{N} X_i = X_1 + X_2 + X_3 + \cdots + X_N$$

is the sum of all of the terms in the universe X. We read this as "The summation of X_i where i assumes all integral values from 1 to N, inclusive." The subscript i is called the *summation index*. Let

$$X_1 = 2, X_2 = 4, X_3 = 8, X_4 = 16$$

Then

$$\sum_{i=1}^{4} X_i = X_1 + X_2 + X_3 + X_4 = 2 + 4 + 8 + 16 = 30$$

The summation index functions like other indices. It points at the items under consideration. Let

$$N = 5, X_1 = -4, X_2 - 3, X_3 = 0, X_4 = -2, \text{ and } X_5 = 1$$

Then

$$\sum_{i=2}^{N} X_i = X_2 + X_3 + \cdots + X_N = 3 + 0 + (-2) + 1 = 2$$

EXAMPLES Given $X_1 = 3, X_2 = -4, X_3 = 6, X_4 = 7$,

find 1. $\sum_{i=1}^{4} (X_i)^2$ 2. $\sum_{i=2}^{3} (X_i - 3X_i^2)$ 3. $\sum_{i=1}^{4} (X_i - 2)^2$ 4. $\left(\sum_{i=3}^{4} X_i \right)^2$

1. $\sum_{i=1}^{4} (X_i)^2 = (X_1)^2 + (X_2)^2 + (X_3)^2 + (X_4)^2$

$$= 9 + 16 + 36 + 49$$
$$= 110$$

2. $\sum_{i=2}^{3} (X_i - 3X_i^2) = (X_2 - 3X_2^2) + (X_3 - 3X_3^2)$

$$= (-4 - 48) + (6 - 108)$$
$$= -154$$

3. $\sum_{i=1}^{4} (X_i - 2)^2 = (X_1 - 2)^2 + (X_2 - 2)^2 + (X_3 - 2)^2 + (X_4 - 2)^2$

$$= (3 - 2)^2 + (-4 - 2)^2 + (6 - 2)^2 + (7 - 2)^2$$
$$= 78$$

4. $\left(\sum_{i=3}^{4} X_i \right)^2 = (X_3 + X_4)^2$

$$= (6 + 7)^2$$
$$= 169$$

Exercises 12-2

1. Express the following in summation notation.
   a. $X_1 + X_2 + X_3 + \cdots + X_N$
   b. $(X_2 + 2X_2) + (X_3 + 2X_3)$
   c. $(X_1 - p)^3 + (X_2 - p)^3 + \cdots + (X_N - p)^3$
   d. $(cX_2)^2 + (cX_3)^2 + (cX_4)^2$
   e. $(X_5 + X_6)^2$

2. Write out all the terms for the following.

a. $\displaystyle\sum_{j=2}^{N} X_j$

b. $\displaystyle\sum_{i=1}^{4} X_i^2$

c. $\displaystyle\left(\sum_{i=5}^{6} Y_i - 2\right)^2$

d. $\displaystyle\sum_{i=2}^{3} (X_i + 2X_i)X_i$

e. $\displaystyle\sum_{k=1}^{N} (aX_k + b)$

3. If $X_1 = 2$, $X_2 = -3$, $X_3 = 0$, $X_4 = 9$, and $c = 4$, find the numerical values for the following.

a. $\displaystyle\sum_{i=1}^{4} X_i$

b. $\displaystyle\sum_{j=2}^{3} (X_j + c)$

c. $\displaystyle\left[\sum_{i=1}^{2} (2X_i - c)\right]^2$

d. $\displaystyle\sum_{i=3}^{4} (X_i^2 - 2X_i)$

e. $\displaystyle\sum_{i=1}^{4} (cX_i)$

4. Does

$$\sum_{i=1}^{2} (X_i)^2 = \left(\sum_{i=1}^{2} X_i\right)^2 .$$

Scores

A score is a count or tally and suggests order and relationship to other scores, i.e., the highest score, the lowest score, or the average score. A single score can be described as a raw score or single variate. A grade of 85 on an examination, for example, can be termed a raw score. Assume that a class of ten made the following raw scores: 63, 65, 73, 75, 81, 83, 85, 91, 95, 100. These scores, as listed, can be described as *ungrouped*. If we listed scores by arranging them as follows

Class	Frequency
61–70	2
71–80	2
81–90	3
91–100	3

we would be discussing *grouped scores*.

The remainder of this chapter contains a discussion of the measures of central tendency and dispersion for ungrouped and grouped scores.

12-3   UNGROUPED SCORES

Measures of central tendency are measurements that indicate the center of a distribution of scores. In this chapter we will be concerned about the mean, the median, and the range.

12-3.1   The Median

Definition 12-5

The *median* is that number on a scale which divides the number of scores into two equal parts.

EXAMPLE 1   Assume the following to be the variates or raw scores on an aptitude test made by nine job applicants: 93, 90, 88, 86, 85, 77, 68, 64, 60. The median is the raw score 85, as there are four scores greater than 85 and four scores less than 85.

If the number of raw scores were an even number, such as 10, we could compute the median as

$$\frac{X_{N/2} + (X_{N/2+1})}{2} \quad \text{or} \quad \frac{X_5 + X_6}{2} \quad (N = 10 \text{ in this example.})$$

EXAMPLE 2   Assuming the ten raw scores to be

$$X_1 = 93, X_2 = 90, X_3 = 88, X_4 = 86, X_5 = 85, X_6 = 77, X_7 = 68,$$
$$X_8 = 64, X_9 = 60, X_{10} = 40$$

then

$$\frac{85 + 77}{2} = \frac{162}{2} = 81$$

The median in this case is 81. It should be noted that the median of 81 is *not* one of the ten raw scores, but the mean of the two middle scores.

12-3.2   The Mean

Definition 12-6

The *mean* (or arithmetic mean) denoted by m for a set of N variates is defined as

$$m = \frac{X_1 + X_2 + X_3 + \cdots + X_N}{N} \quad \text{or} \quad m = \frac{\sum\limits_{i=1}^{N} X_i}{N}$$

Considering our sample of nine raw scores presented in Example 1 (where $N = 9$), we can compute m as follows:

$$m = \frac{\sum_{i=1}^{N} X_i}{N} = 79$$

or

$$\frac{93 + 90 + 88 + 86 + 85 + 77 + 68 + 64 + 60}{9} = 79$$

If several of the raw scores were 40 or below, they would have noticeably lowered the test mean. Similarly, if several of the test scores were above 93, they would have significantly raised the mean. In such situations the median has a significance as a measure of central tendency.

12-3.3 The Range

Definition 12-7

For a set of N numbers, the *range* is a computed distance between the largest and smallest numbers, i.e., the difference.

The range of the scores in the last example is:

median

93, 90, 88, 86, 85, 77, 68, 64, 60

range $= 93 - 60 = 33$

Or, stated in another manner, the scores range from 60 to 93. Note that within the range there is a scattering of scores about a central point, such as the mean or median.

12-3.4 Measure of Dispersion

The arithmetic mean is a helpful measure of a central tendency, and the median is useful when extremely high or low scores occur as noted earlier. The range gives some measure of the dispersion of scores about the mean. However, none of these three measurements indicates the deviations from the mean or the distribution of the variates either at the extremes of the range or at the center.

Consider the following three sets of raw scores:

EXAMPLE 1   20, 50, 80, 90, 100
median = 80
mean   = 68
range = 80

EXAMPLE 2
$$30, 45, 80, 85, 90$$
$$\text{median} = 80$$
$$\text{mean} \ \ = 66$$
$$\text{range} \ \ = 60$$

EXAMPLE 3
$$15, 65, 75, 80, 95$$
$$\text{median} = 75$$
$$\text{mean} \ \ = 66$$
$$\text{range} \ \ = 80$$

Examples 1 and 2 have the same median. Examples 1 and 3 have the same range. Examples 2 and 3 have the same mean. Examination of

—examples 1 and 2 shows that the median does not describe these two sets of raw scores equally well.
—examples 1 and 3 shows that the range does not describe these two sets of raw scores equally well.
—examples 2 and 3 shows that the mean does not describe these two sets of raw scores equally well.

We need, therefore, some method to describe the spread or dispersion of a set of raw scores about their mean. This measure of the spread of all scores is called the *standard deviation.*

Before studying the standard deviation, we must first consider the *deviation* and the *variance,* which are required to compute standard deviation.

Definition 12-8

A *deviation* of a score X_i is the difference between X_i and m. A deviation is denoted as d, where $d = X_i - m$.

In Example 1 (page 266) this works out for X_5 as
$$d = X_5 - m$$
$$d = 100 - 68$$
$$d = 32$$

Note that in Example 1, $\Sigma (X - m) = 0$:

$$\sum_{i=1}^{5} (X_i - m) = (X_1 - m) + (X_2 - m) + (X_3 - m)$$
$$+ (X_4 - m) + (X_5 - m)$$
$$= (20 - 68) + (50 - 68) + (80 - 68)$$
$$+ (90 - 68) + (100 - 68)$$
$$= (-48) + (-18) + (12) + (22) + (32)$$
$$= 0$$

As an exercise, show that $\Sigma (X - m) = 0$ for examples 2 and 3.

Definition 12-9

A *variance* is the sum of the deviations squared divided by N, i.e.

$$\text{Variance} = \frac{\sum_{i=1}^{N} (X_i - m)^2}{N}$$

and is denoted by σ^2.

Examine Table 12-1.

Table 12-1

Score	f	$X_i - m = d$		d^2
93	1	93 − 79 =	14	$(14)^2$
90	1	90 − 79 =	11	$(11)^2$
88	1	88 − 79 =	9	$(9)^2$
86	1	86 − 79 =	7	$(7)^2$
85	1	85 − 79 =	6	$(6)^2$
77	1	77 − 79 =	−2	$(-2)^2$
68	1	68 − 79 =	−11	$(-11)^2$
64	1	64 − 79 =	−15	$(-15)^2$
60	1	60 − 79 =	−19	$(-19)^2$

Scores are found in the left column in descending order. (They may also be presented in ascending order.) The second column from the left contains the frequency (f) for each score. (Since in this example we are using raw scores, only one frequency can be written for each score achieved.) Marking off the deviations from the mean gives us the next column, d.

At this point we wish to find the average dispersion of scores from a central point. But notice that, in adding up the column under d, negative and positive values cancel each other out. To avoid + and − designations, we square the deviation and take the mean of the squares. This may be represented by the formula

$$\frac{\sum_{i=1}^{N} d_i^2}{N}$$

In our example this works out as $\frac{1194}{9}$ = 132.7. But 132.7 is a measure of variance as a square. To find this variance as a linear measurement, we take the square root of 132.7, or $\sqrt{132.7}$ = 11.52.

Definition 12-10

The square root of the variance is called the *standard deviation*.

The symbol σ is used to denote the *standard deviation*. The following is the formula to find the standard deviation for ungrouped scores and illustrates taking the square root of the sum of the deviations squared, divided by N, which we have called the variance.

$$\sigma = \sqrt{\frac{\sum_{i=1}^{N} d^2}{N}}$$

Other variations of this formula are possible. Two which are easy to code for computer programs are

$$\sigma = \sqrt{\frac{N \sum X_i^2 - (\sum X_i)^2}{n(n - 1)}}$$

where n represents a large group of variates, and

$$\sigma = \sqrt{\frac{\sum_{i=1}^{N} d_i^2}{N - 1}}$$

where N represents a small number of variates.

12-4 GROUPED SCORES

Table 12-2 contains the deposits made in 100 banks for one year. The deposits are expressed in millions of dollars ($50.5 = \$50,500,000$). The amounts have been placed in descending numerical order.

Table 12-2

Deposits

50.5	46.8	42.2	39.2	38.4	36.2	32.1	27.1	21.4	15.3
50.3	46.6	41.6	39.1	38.3	36.1	31.6	25.9	21.3	14.4
50.0	45.5	41.6	39.1	38.3	36.1	31.5	25.7	20.3	13.9
48.6	45.0	41.5	39.0	38.2	35.6	30.7	25.6	20.1	12.7
48.5	44.8	41.3	39.0	38.0	35.6	30.6	25.0	19.6	11.7
47.3	44.7	41.1	38.8	37.9	34.8	28.4	24.7	18.4	10.2
47.1	44.5	40.7	38.8	37.9	34.2	28.3	24.6	17.9	9.6
47.0	43.7	40.6	38.8	37.7	32.9	27.5	23.2	16.2	7.1
46.9	43.2	40.0	38.6	37.4	32.4	27.4	23.0	15.6	7.0
46.9	42.3	39.8	38.5	37.3	32.3	27.2	22.5	15.5	5.5

This arrangement of the distribution of variates is cluttered and bulky. Table 12-3 shows the distribution of these variates in a more compact format that is easier to visualize. The data now can be easily manipulated and measured because the interval of measurement includes not just one variate, but a group of variates. We will refer to such a group of scores as a *class*. Each class has its own upper and lower boundary and contains those values that properly belong in their respective class intervals.

Classes are arranged in ascending numerical order in the first column in Table 12-3.

Table 12-3

Class Boundaries (in millions)	Class Midpoints	Tallied Frequencies	Class Frequency	Cumulative Frequency
5.5–10.5	8	⊮	5	5
10.5–15.5	13	⊮	5	10
15.5–20.5	18	⊮ |||	8	18
20.5–25.5	23	⊮ |||	8	26
25.5–30.5	28	⊮ ||||	9	35
30.5–35.5	33	⊮ ⊮	10	45
35.5–40.5	38	⊮ ⊮ ⊮ ⊮ ⊮ ||	27	72
40.5–45.5	43	⊮ ⊮ ⊮	15	87
45.5–50.5	48	⊮ ⊮ |||	13	100

Note that the last column, cumulative frequency, is computed by adding up the class frequencies row by row. For example, for the first class the cumulative frequency is 5. For the second class the cumulative frequency is 5 + 5 = 10. For the third class the cumulative frequency is 10 + 8 = 18, and so on. The cumulative frequency is shown graphically in Fig. 12-12.

The range for grouped scores is the computed distance between the upper boundary of the highest class and the lower boundary of the lowest class. As an exercise, compute the range for Table 12-3, referring to Definition 12-7 if necessary.

The nine class boundaries in the left-hand column include all of the dollar amounts found in Table 12-3. The *class midpoint* is halfway between the upper and lower boundary. The number of data items falling within any class is called the *class frequency*. In this chapter any data item falling on a class boundary shall be placed in the higher class.* (Class boundaries could be specified to the hundredths place to avoid variates falling on them.)

In this example, the class interval was determined by taking the difference of the endpoints of the range (50.5 − 5.5 = 45.0), dividing by 10 ($\frac{45}{10}$ = 4.5), and rounding to the nearest integer.

It is sometimes useful to display grouped scores in graphic form. Graphs and histograms display information in a compact form and differences in magnitude are easy to recognize. (See Figure 12-10.)

The term *mode of the distribution* is used to describe that part of the scale at which more variates are found than at any other place. It is a

---

* Except for 50.5, which is the greatest variate in Table 12-2.

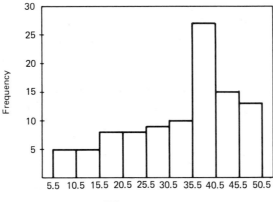

FIGURE 12-10

characteristic of the concentration of scores that cannot be expressed by any other measure of central tendency. In Fig. 12-10 we can see that the mode of the distribution occurs in the class bounded by 35.5 and 40.5.

In grouped scores or variates, a true mode requires several computations and is used to express the place (or places) on a scale where frequencies occur most often.

If we rotate Table 12-3 90° counterclockwise, we are able to create a histogram (Fig. 12-10) or vertical bar graph, with class boundaries on the horizontal axis and frequencies on the vertical axis. By connecting class midpoints (second column in Table 12-3), we can then draw the frequency polygon shown in Fig. 12-11. The broken lines in the frequency polygon of Fig. 12-11 connect the midpoints of the tops of the rectangles of Fig. 12-10.

To illustrate graphically the accumulation of monies deposited in the one hundred banks, we create a cumulative frequency polygon. Take the first column from Table 12-3 as our horizontal axis, and use the last column in Table 12-3 as our vertical axis. (See Fig. 12-12.) Upper-class

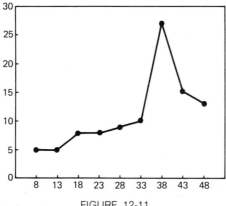

FIGURE 12-11

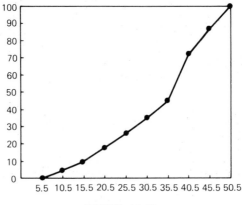

FIGURE 12-12

boundaries are indicated by points on the graph, and these points are connected by line segments.

12-4.1 The Mean

One method of calculating the mean for grouped scores assumes that the data items in each class are uniformly distributed about that class midpoint, where

$$m = \frac{\sum_{i=1}^{K} X_i f_i}{N}$$

for

N = the total of all the frequencies for all classes
X_i = the ith midpoint
f_i = the ith frequency
K = the number of classes
m = the mean

For the data given in Table 12-4, the mean for the grouped scores (Xf column) can be computed as

$$\begin{aligned}
&\frac{\begin{aligned}8(5) + 13(5) + 18(8) + 23(8) + 28(9)\\ + 33(10) + 38(27) + 43(15) + 48(13)\end{aligned}}{100}\\[2mm]
&= \frac{40 + 65 + 144 + 184 + 252 + 330 + 1026 + 645 + 624}{100}\\[2mm]
&= \frac{3310}{100}\\[2mm]
&= 33.10
\end{aligned}$$

Table 12-4

Class Boundaries	X	f	Xf
5.5–10.5	8	5	40
10.5–15.5	13	5	65
15.5–20.5	18	8	144
20.5–25.5	23	8	184
25.5–30.5	28	9	252
30.5–35.5	33	10	330
35.5–40.5	38	27	1026
40.5–45.5	43	15	645
45.5–50.5	48	13	624

12-4.2 The Median

It was a simple operation to find the median for ungrouped scores. However, to find the median for grouped scores requires more computation. In grouped scores the median is found in the *median class*, or in the lowest class in which the cumulative frequency is greater than $N/2$.

Let

N = the total of all the frequencies for all classes
L = the mean of the lower boundary of the median class and the upper boundary of the next lower class
c = the difference between two adjacent class midpoints
fm = the frequency of the median class
F_c = the cumulative frequency for the class just below the median class (specifies all the variates up to but not including L)
M_d = the median for grouped scores

Assume that the data items in the median class are distributed uniformly throughout the interval. (Refer to Table 12-3; the last column implies that the median class is $35.5 - 40.5$.) Then

$$c\left(\frac{N/2 - F_c}{fm}\right)$$

is the distance from L to the median, or the correction added to L to find the median. Therefore

$$M_d = L + c\left(\frac{N/2 - F_c}{fm}\right)$$

$$M_d = 35.5 + 5\left(\frac{100/2 - 45}{27}\right)$$

$$= 35.5 + 5(.185)$$

$$= 36.425$$

Note that when N is even, the formula for the median considers the first of the two middle numbers, not their mean. (See page 265.) This is satisfactory, since a frequency distribution of grouped scores will usually contain enough variates so that the middle numbers differ by very little.

Consider that the number of variates less than L is subtracted from half of the total number of variates. Divide this difference by the frequency of the median class. The resulting number is equivalent to the number derived by subtracting the mean of the lower boundary of the median class and the upper boundary of the next lower class from the median and dividing this difference by the difference between two adjacent class midpoints, or

$$\frac{N/2 - F_c}{fm} = \frac{M_d - L}{c}$$

12-4.3 Standard Deviation

The standard deviation for grouped scores can be computed easily if we drop the fourth column in Table 12-4 and add four new columns to Table 12-4. (Note that $|X - m|$ means the absolute value of $X - m$.)

Class Boundaries	X	f	$|X - m|$	$|X - m|f$	$(X - m)^2$	$(X - m)^2f$
5.5–10.5	8	5	25.1	125.5	630.01	3150.05
10.5–15.5	13	5	20.1	100.5	404.01	2020.05
15.5–20.5	18	8	15.1	120.8	228.01	1824.08
20.5–25.5	23	8	10.1	80.8	102.01	816.08
25.5–30.5	28	9	5.1	45.9	26.01	234.09
30.5–35.5	33	10	.1	1.0	.01	.10
35.5–40.5	38	27	4.9	132.3	24.01	648.27
40.5–45.5	43	15	9.9	148.5	98.01	1470.15
45.5–50.5	48	13	14.9	193.7	222.01	2886.13

By modifying the formula used earlier for ungrouped scores, we obtain the following formula for grouped scores (where K = the number of classes):

$$\sigma_g = \sqrt{\frac{\sum_{i=1}^{K} (X_i - m)^2 f_i}{N}}$$

σ_g is the symbol used to denote the standard deviation for grouped scores, and K = the number of classes. Thus, for the data in our table

$$\sigma_g = \sqrt{\frac{13049}{100}}$$
$$= 11.423$$

Exercises 12-3

1. Given the set of numbers

$$4.9, 1.5, 3.9, 0.1, 6.2, 8.4, 9.5, 7.8, 2.3$$

find the following to two decimal places.

a. The median
b. The mean
c. The standard deviation

2. Given

Class Boundaries	Frequencies
6.5– 9.5	3
9.5–12.5	7
12.5–15.5	8
15.5–18.5	17
18.5–21.5	24
21.5–24.5	18
24.5–27.5	11
27.5–30.5	4

Find

a. The mean
b. The median class
c. The median
d. The standard deviation

3. Describe the following Fortran statements.

a. SUM = A + B + C + D + E
b. DIFF = (XN/2) − FC

(Note: The parentheses denote to the computer the operation to be performed first. If there are nested parentheses, the terms in the inner-most parentheses are computed first.)

c. PAY = ((HRS − 40.0) * RATE * 1.5) + 40.0 * RATE
d. RATE = PAY/HRS

4. Code a Fortran statement to find the averages of the variables A, B, C, D, and E in problem 3a.

Normal Maximum Airport Temperature From Selected Cities in Fahrenheit Degrees
(Based on Standard 30-year Period, 1931 to 1960.)

Station	Jan.	Feb.	Mar.	Apr.	May	Jun.	Jul.	Aug.	Sept.	Oct.	Nov.	Dec.
Mobile	62.3	64.7	70.3	77.5	85.9	91.4	92.0	91.2	87.4	80.3	69.6	63.9
Juneau	30.1	32.1	36.5	45.4	53.6	60.8	62.7	61.5	55.2	46.5	39.2	32.7
Little Rock	50.6	54.6	62.7	73.5	81.5	89.7	92.7	92.4	86.3	76.0	61.3	52.1
Los Angeles	63.8	63.7	65.0	66.9	68.7	71.1	75.9	75.4	75.8	73.0	71.0	66.5
Sacramento	53.2	58.6	64.8	71.4	78.2	86.5	93.4	91.9	88.2	77.6	64.2	54.6
San Francisco	55.8	58.6	60.7	61.9	63.4	65.0	64.3	64.9	68.9	68.3	63.7	57.5
Denver	41.1	44.6	49.9	60.5	70.5	82.0	88.4	86.8	79.0	66.6	51.7	45.2
Hartford	34.7	36.0	45.3	59.6	72.0	80.5	85.0	82.7	74.7	64.8	50.9	37.6
Wilmington	41.3	42.4	50.5	62.5	73.4	81.8	86.2	84.2	77.9	67.3	55.1	43.5
Washington	44.3	46.1	53.8	65.8	75.5	83.4	87.0	85.0	78.6	68.3	56.5	45.6

5. Write Fortran statements to find the following.

   a. The mean for grouped scores in four classes, where

        XM = the mean
        XI = the ith midpoint
        FI = the ith frequency
        XK = the number of classes
        XN = the total numbers of frequencies for all classes

   b. The median for grouped scores. Create your own variable names, but remember that letters $I - N$ are used as the first letter in a variable name for integers only.

   c. The standard deviation for

      1. ungrouped scores
      2. grouped scores.

6. Given: Normal Monthly Maximum Airport Temperature from Selected Cities in Fahrenheit Degrees (p. 276).

   a. From the temperature data given, complete a table using the headings

     Class

     Boundaries   X   f   Xf $|X - m|$ $|X - m|f$ $(X - m)^2$ $(X - m)^2 f$

     (A method for determining class boundaries for grouped scores may be found on page 270 of this chapter.)

   b. Find:

      1. The range
      2. The mean
      3. The median
      4. The standard deviation

   c. Draw the appropriate histogram.
   d. Draw the frequency polygon.
   e. Draw the cumulative frequency polygon.

13

PROGRAMMING CONSIDERATIONS

Computer programming frequently involves several types of problems:

1. Arithmetic operations of addition, subtraction, multiplication, and division
2. Overflow/truncating
3. Rounding
4. Signed numbers
5. Input/Output checks
6. Table look-up
7. Computing elapsed time/retention dates
8. Binary search

13-1 ARITHMETIC OPERATIONS

Arithmetic in computers is usually performed in registers and fields. Because the size of registers is not typically under the control of the programmer, but fields are, we will only be concerned here with fields.

Definition 13-1

A *field* is one position or group of consecutive positions reserved for data on punched cards, magnetic tapes, magnetic disks, or in computer storage.

Assume that the following are positions reserved for a social security number:

A |_|_|_|_|_|_|_|_|_|_|_|

Since each social security number contains nine digits, the field we have labeled A contains enough positions to accept each of the necessary digits. An example is

|0|9|6|1|2|9|5|6|7|

(with the editing dashes omitted).

13-1.1 Addition

Consider the following situation for Assembler Language. Assume that we are adding the information from two fields called Z and Y into a third field called X. Since each field, including the result field, is three positions long, we will not be able to accommodate any overflow or "carry" out of the high-order position. If Y = 111 and Z = 888, there is no overflow; but if Y = 112 and Z = 888, then

```
  Y    |1|1|2|
+ Z    |8|8|8|
  X   1|0|0|0|
```

X would contain all zeros in this case because the carry or overflow out of the high-order position of X would be truncated or lost.

To prevent the truncation of high-order digits in an overflow situation, the programmer may
1. Provide result fields which are long enough to accommodate sums (and products), or
2. Test for an overflow condition and instruct the machine to take remedial action as shown in the flowchart for a typical overflow problem, Fig. 13-1.

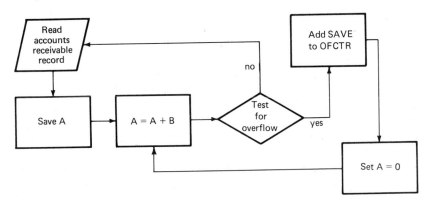

FIGURE 13-1

In Fig. 13-1 if B were found on an input tape and A were a location inside computer storage used to accumulate totals, then reading hundreds of tape records presents an overflow problem to the programmer.

The flowchart in Fig. 13-1 illustrates the logic used to *save* totals when an overflow condition can cause truncation out of the high-order position. When the overflow condition occurs, the SAVE field is added to OFCTR, an overflow counter. The 360 Assembler Language coding for this method of overflow handling is given below.

```
READ    GET   TAPE, INAREA   READ A TAPE RECORD
        MVC   SAVE, A         SAVE DATA
ADD     AP    A, B            ADD B TO A
        BC    1, OVFLO        IF OVERFLOW GO TO OVFLO
        BC    15, READ        GO TO READ
OVFLO   AP    OFCTR, SAVE     ADD SAVE TO OVERFLOW COUNTER
        SP    A, A            SET A TO ZERO
        BC    15, ADD         GO TO ADD
```

It is also possible to *subtract out* the addend that caused the overflow condition, as we shall see in the following Fortran example. Before coding the example, however, we note that for one frequently used Fortran compiler, the highest possible positive integer is 32767 or $(2^{15} - 1)$, and the lowest possible negative integer is $-(2^{15})$.

Study the following:

```
          INTEGER X,Y
          X = 32767
          WRITE(3,100)X
          Y = 9
     2    X = X + Y
          WRITE(3,100)X
   100    FORMAT(1X,I6)
          IF (X) 3,1,1
     1    GO TO 4
     3    X = X - Y
          WRITE(3,100)X
          OFLO = OFLO + X
          X = 0
          GO TO 2
     4    CALL EXIT
          END
           32767
         - 32760
           32767
               9
```

A general description of the activities in this Fortran program follows.

X is declared to be an integer, initialized to 32767, and printed out at the bottom of the listing.

Y is initialized to 9 and added to X, causing a negative result of − 32760. This result occurred in the following manner:

$$
\begin{array}{rcl}
32767 + 1 & = & -32768 \\
+ 1 & = & -32767 \\
+ 1 & = & -32766 \\
+ 1 & = & -32765 \\
+ 1 & = & -32764 \\
+ 1 & = & -32763 \\
+ 1 & = & -32762 \\
+ 1 & = & -32761 \\
+ 1 & = & -32760 \\
\end{array}
$$

When 9 was added to 32767, X went first to the lowest possible negative number, − 32768, and then increased in value to − 32760.

The statement

$$IF (X) 3, 1, 1$$

tested X and found it to be negative, and program control was given to statement 3. At statement 3, Y was subtracted from X. Next, X was added to an overflow counter, and then set to zero. Program control was given to statement 2, where Y could be added to X yielding 9.

If we require several overflow counters, we change statement 3 to read:

$$3 \text{ GO TO } (10, 20, 30), N$$

This statement is the Fortran computed GO TO, which transfers program control to the statement number in the parenthesized list in an order corresponding to the value of N. If $N = 1$, program control is given to the first statement in the list. If $N = 2$, program control is given to the second statement in the list, and so on.

N can be initialized to 1 at the beginning of the program. Statements 10, 20, and 30 may then give program control for instructions to save X in three overflow areas. For example:

```
10  X = X − Y
    OVA = X
    N = N + 1
    X = 0
    GO TO 2
    _____

    _____  other statements in the program

    _____
```

```
20   X = X – Y
     OVB = X
     N = N + 1
     X = 0
     GO TO 2

     _____
     _____      other statements in the program
     _____

30   X = X – Y
     OVC = X
     N = N + 1
     X = 0
     GO TO 2
```

Overflow conditions in PL/I and Cobol are handled by ON conditions which are found in appropriate reference manuals.

13-1.2 Subtraction

It is a simple programming matter to make the difference or result field the same size as that of the minuend, thus:

$$XXXXXX \quad \text{Minuend}$$
$$\underline{XXX} \quad \text{Subtrahend}$$
$$XXXXXX \quad \text{Difference}$$

However, tests must always be made after subtraction in the event that negative results are unwanted. Consider a payroll situation where, due to sickness or other reasons, an employee missed a great amount of time and pay. Such voluntary deductions as union dues, savings plans, and the United Fund may actually amount to more than the money earned for the period. In this situation his net pay may be smaller than his exemption allowance. Tax calculations might be made on a negative amount. In such payroll situations the program must test for negative numbers before tax calculations. Thus, in

IF (PAY – EXEMT) 1, 1, 2

statement 1 should contain an instruction to bypass tax calculations.

13-1.3 Multiplication

When two numbers are multiplied together, the resulting number of digits in the product depends upon the magnitude of the two numbers. Consider, for example,

PAY = RATE * HOURS

RATE is a constant stored internally in the computer, and the field HOURS contains data items on a deck of employee payroll time cards. The result field is PAY.

As each card is read and processed, the programmer has no way of knowing exactly what the values of HOURS are going to be for each employee, but if he knows their field length, he can determine the range of values. To accommodate the result of the largest possible magnitude, the length of the product field must be equal to the sum of the lengths of the multiplier and the multiplicand. If

the multiplicand is	999	3 digits
and the multiplier is	99	2 digits
then the product is	98901	5 digits

It is a good practice to use 9's as multiplier and multiplicand in determining the length of the result field. In the example that follows, if the product is shorter than six digits, the high-order position(s) will be filled with zeros:

2451	4 digits
12	2 digits
029412	6 digits

During arithmetic operations in a computer, the decimal point is an *assumed* point in the fields being operated on. For example,

$$\boxed{1\ |\ 2\ |\ 3} \quad \text{means} \quad 1.23$$

$$\boxed{7\ |\ 3\ |\ 2\ |\ 1\ |\ 0} \quad \text{means} \quad 7.3210$$

The caret ($\wedge$) indicates the assumed decimal point.

In assembler languages, the programmer must keep track of the decimal point. In higher-level languages such as Fortran, *PL/I*, Cobol, and Basic, the computer will perform this function if properly instructed by the programmer. To avoid truncating of low-order decimal fractions, care must be taken to define result fields with the desired number of positions to the right of the decimal point. For example, in

9.99
9.9
98.901

because there are two decimal positions in the multiplicand and one in the multiplier, there are three resulting decimal fraction positions. In some computer languages, such as *PL/I*, the product field in the computer should contain these three positions or truncation will cause the loss of the rightmost decimal position. Decimal fractions can be rounded off; this is discussed on page 286.

13-1.4 Division

A division operation can be expressed in terms of divisor, dividend, quotient, and remainder, where

$$\frac{dividend}{divisor} = quotient + remainder$$

Since division by zero is not permitted on the computer, all divisors should be tested before the division operation is executed.

IF (DIVSR.EQ.0.00) GO TO 2

is a Fortran statement that will cause program control to be transferred to statement 2 if DIVSR has a value of zero. Statement 2 should contain remedial activity or another instruction to bypass the division operation.

Zero divisors in Cobol can be trapped very simply by means of the statement

IF DIVISOR = 0 GO TO BY-PASS.

where BY-PASS is the name of a paragraph containing the instruction(s) to perform some corrective action.

Because the smallest possible number of positions for any divisor field is one, it follows that the largest possible number of digits in the quotient must equal the number of digits in the dividend.

EXAMPLE 1 $\dfrac{2222}{1} = 2222$

EXAMPLE 2 $\dfrac{1206}{6} = 0201$

Note the padding or use of the high-order zero to denote a four-position result field.

EXAMPLE 3 $\dfrac{88.888}{2} = 44.444$

Note that .444 is *not* a remainder.

EXAMPLE 4 $\dfrac{99.99}{.3} = 333.3$

The results in the last two examples contain decimal fractions. The computer performs arithmetic with *assumed* decimal points (printed decimal points are edited into result fields by the programmer). The rule for the placement of the decimal point in division is the same one we use when writing out the problem ourselves. The number of decimal positions in the quotient is equal to the number of decimal positions in the dividend minus the number of decimal positions in the divisor. This follows from the rule for multiplication, since the divisor times the quotient equals the dividend.

13-1.5 Remainders

Remainders are often the unusable results of a divide operation and contain the same number of positions as divisors. Consider:

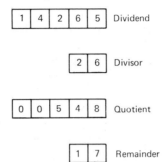

To determine our result to two decimal fractions, we multiply our dividend by 10^2 or 100, thus:

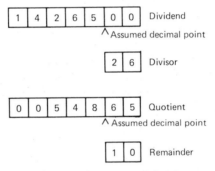

Using X's for placeholders in our operand fields, note that

$$\frac{X.XXX}{.X} \text{ is equivalent to } \frac{XX.XX}{X.}$$

In assembler languages some programmers multiply by an appropriate power of 100 to remove decimal fractions in the dividend. (In this last case we multiplied by 10^1.) This presents no problem to the computer, since the actual decimal point is never present in an operand field. After the arithmetic operation has taken place, the programmer issues an edit instruction inserting the decimal point in the appropriate place.

13-2 ROUNDING

Results of arithmetic operations often include more decimal places than are required in the final result. This is often true when computing interest rates and tax calculations where the final answer is usually expressed in dollars and cents. Unwanted decimal positions must be rounded to the nearest cent; for example, $3.546 becomes $3.55 after rounding and − $3.546 becomes − $3.55 after rounding.

13-2.1　Positive and Negative Numbers

Conventionally, rounding has been performed by adding 5 in the column to the right of the number to be rounded and truncating the unnecessary digit(s).

	a.	− 147	b.	14296	c.	191
	+ −	5	+	50	+	5
		− 152		14346		196

In Example a, we rounded to the nearest two digits. In Example b, we rounded to the nearest three digits. In Example a, we added in a negative 5 because the number to be rounded was negative. In Example c, there is no carry because the digit to be eliminated is less than 5.

If we know that a field named X has five positions, including dollars and cents, we can round to the nearest positive or negative dollar amount. The following coding illustrates this procedure in Fortran.

```
     ROUND = .50
     IF (X)20,10,10
10  RESLT = X + ROUND

     _____

     _____    other statements in program

     _____

20 ROUND = − ROUND
     GO TO 10
```

A different procedure may be followed in assembler languages, where programmers can actually access the digit to be rounded and the digit(s) to be truncated. Instead of adding + 5 or − 5 to the digit(s) to be dropped, the digit(s) under consideration may be added algebraically to itself or themselves.

EXAMPLES

	a.	2.34 +	b.	4.78 −
		+ .04 +		+ .08 −
		2.38 +		4.86 −
		2.3 +		4.8 −
	c.	24.5673 +	d.	18.4321 −
		+ .0073 +		+ .4321 −
		24.5746 +		18.8642 −
		24.57 +		18 −

(Note: the sign of the number appears next to the low-order position.)

13-3 INPUT/OUTPUT CHECKS

All input and output must be checked for validity. Sometimes this is done visually by inspection of input cards or printed reports.

13-3.1 Input Verification

All input must be valid, or processing and output are meaningless. Keyboard operators, clerks, bookkeepers, accountants, tellers, salesmen, and customers all make errors when writing numbers. A method of verifying identification or payroll numbers uses check digits. Gasoline credit cards contain examples of check digits.

There are various methods of determining a check digit, and frequently much is left to the ingenuity of the programmer. Two examples are given here:

EXAMPLE 1 Let n = an account number; then $n/m = q + r$, where r becomes the check digit. This check digit is annexed to the low-order position of the account number.
For example let

$$n = 4587$$
$$m = 6$$

Then

$$\frac{4587}{6} = 764 + 3/6$$

Annexing r to the low-order position of n, we get the account number with its check digit, or 45873.

EXAMPLE 2 Let

$$n = \text{a payroll number}$$
$$m = \text{a one-digit constant}$$

Subtract m from all but the first two digits in n. Add the digits in the difference and cast out nines until the last digit is less than or equal to nine. Annex this digit to the low-order digit of the payroll number as a check digit.

As an example, assume that the employees' social security numbers were used for payroll numbers (n) and that $m = 2$. Then

$$\begin{array}{r} 123456789 \\ -2222222 \\ \hline 121234567 \end{array}$$

$$1 + 2 + 1 + 2 + 3 + 4 + 5 + 6 + 7$$

Casting out nines means $1 + 2 + 1 + \cancel{2} + \cancel{3} + 4 + \cancel{5} + \cancel{6} + \cancel{7} = 4$. The check digit, 4, is annexed to the social security number after the low-order digit, thus: 1234567894.

13-3.2 Output Checks

Errors must be corrected before they spread throughout a business system. Catching these errors cannot always be done visually. When data items are found on magnetic tapes or disks for example, other methods are used.

HASH TOTALS

Consider item numbers in an inventory system. Each item number is placed at the beginning of a record in a tape file of inventory records. (See Fig. 13-2.)

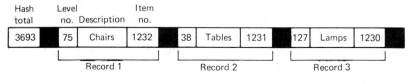

FIGURE 13-2

Each record in the inventory file contains an item number, a description, and a level number, i.e., a number denoting the amount on hand. Inventory levels are checked periodically to determine whether they are overstocked or understocked. To be certain that each record has been interrogated, a "hash" total is taken. This means adding up each of the item numbers each time the tape is read. To make certain that each record is read, the computed hash total is compared to the hash total written at the end of the tape file. If the hash totals agree, each record has been read and processed. If the hash totals are not equal, an error situation exists.

PROOF NUMBERS

Some payroll situations, such as those found in educational and other public institutions, require that the same net pay is given to each employee each week. This means that Federal and State tax amounts and voluntary deductions will always remain constant (unless there is a change in dependents or tax laws). By signing alternately + and −, and by algebraically adding such amounts as gross pay, Federal tax, State tax, United Fund, etc., a unique *proof* number can be assigned to each employee each pay period. This proof number can be used to verify that each employee's check has been correctly processed.

Table 13-1

| + | − | + | − | + | − | |
Employee Number	Gross Pay	Fica Tax	Federal Tax	State Tax	United Fund	Proof Number
110412	960.00	58.85	190.01	40.00	2.00	005096
132164	1040.00	69.16	215.61	42.40	2.00	017559

Note the signs placed above each column. Each of these fields is algebraically added together for each employee to produce a proof number. (Decimal points are ignored.)

CROSSFOOT TOTALS

Suppose that an input file of accounts-receivable cards contains amounts for quarterly sales as shown in Fig. 13-3. Let Q1 represent sales from January through March, Q2 represent sales from April through June, and so on. Suppose it were necessary to

1. Total each account receivable for the year and take a grand total of sales by accounts.
2. Total each quarter sales and take a grand total of sales by quarter.

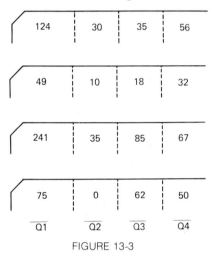

FIGURE 13-3

Since both grand totals must agree, the programmer must instruct the computer to take crossfooted totals, or totals by rows, as shown below.

```
Q1    Q2    Q3    Q4
124 + 30 + 35 + 56 = 245
 49 + 10 + 18 + 32 = 109
241 + 35 + 85 + 67 = 428
 75 +  0 + 62 + 50 = 187
                     969   Grand total of sales by accounts
```

Then the totals are computed by columns:

```
   Q1     Q2     Q3     Q4
  124     30     35     56
   49     10     18     32
  241     35     85     67
+  75   +  0   + 62   + 50
  489     75    200    205 = 969   Grand total of sales by quarter
```

If the two grand totals do not agree, an error condition exists which must be corrected before the output, or total yearly sales, can be sent to the computer's output device.

13-4 TABLE LOOK-UP

Tables are used in many business calculations. Some common examples are

1. State tax tables
2. Federal tax tables
3. Rate tables for utility companies
4. Insurance tables
5. Interest tables

Consider Table 13-2, made up of entries which contain customer numbers (sequenced in ascending order) and balance due amounts.

Table 13-2

	Customer Number	Balance Due ($)
Entry 1	10025	300.00
Entry 2	11620	475.81
Entry 3	20009	300.00
Entry 4	36125	029.00
Entry 5	40201	750.25

Definition 13-2

A table *entry* is a collection of related facts treated as a unit.

Over a period of time the balance due can vary with new purchases and part-payments. However, the number assigned to any customer will not change, so that it can serve as a key or number used to identify the customer's account. This key or identification number is also referred to as the table argument.

Definition 13-3

A table *argument* is a unique number found in an entry in a table. The table argument serves as a key to reference other related facts in the table entry.

Definition 13-4

The table *function* is the information contained in a table entry and referenced by the table argument.

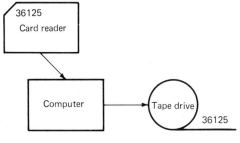

FIGURE 13-4

Each table argument can have one corresponding function, but a function may have one or more table arguments. In Table 13-2, more than one customer could have a balance due of $300.00.

Assume that Table 13-2 exists on a magnetic tape file and that an inquiry was made on customer number 36125 from the card reader. To access account 36125 on tape, we must input that account number from an inquiry device. Using a punched card as the inquiry device, the number 36125 (on the punched card) is called the search argument.

Definition 13-5

A *search argument* is a value given in input data. It is used to locate an equal comparison between itself and the table argument.

The logic used to find the account number on the tape is as follows:

1. Compare the search argument to the table argument.
2. If the comparison finds the two arguments to be equal, branch to a routine in the program to process the function.
3. If the comparison finds the two arguments not to be equal, have we examined each table argument in the tape?
4. If we have not compared every account number to the number on the card, move ahead one entry to compare the next table argument to the search argument.
5. If we have examined every table argument and found unequal comparisons, the search argument is invalid; that is, there is no corresponding entry in the tape, and the inquiry card should be bypassed.

We can flowchart our logic as shown in Fig. 13-5.

_____ **13-5 COMPUTING ELAPSED TIME**

Electronic Data Processing (EDP) files have various retention periods. Tax information must be retained permanently. Accounts

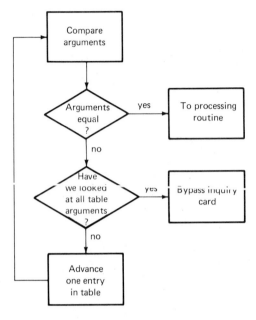

FIGURE 13-5

receivable are updated periodically (monthly or quarterly). Employee year-to-date figures change with each pay period. In addition, calculating the elapsed time between dates is important in determining interest on loans. To help determine elapsed time, a retention date is written on an EDP file.

If we assign each successive day in a year a unique number from 1 to 365, it is a simple matter to determine elapsed time by subtraction. For example:

Day-Number	Calendar Date
71	March 12
− 38	February 7
33	

Using this technique we can quickly determine that 33 days have elapsed between February 7 and March 12.

To create a table of days with each table argument containing the month and day (Jan. 1, Jan. 2, etc.) and each table function containing 001, 002, etc. would be costly in terms of computer storage and table look-up time. Instead, we use a smaller table containing one table argument for each month and a function containing accumulated days. (See Table 13-3.)

If we are interested in 0815, or August 15, we search for 08 and add 15 to 212. August 15 is the 227th day of the year. Assume that a date found on a tape file was 156 and an input card contained the date 0815,

Table 13-3

Argument (month)	Function (accumulated days)
01	000
02	031
03	059
04	090
05	120
06	151
07	181
08	212
09	243
10	273
11	304
12	334

or August 15; the elapsed time can be determined by a table look-up routine and the operations of addition and subtraction:

$$\begin{array}{r} 212 \\ +\ 15 \\ \hline 227 \end{array} \qquad \begin{array}{r} 227 \\ -156 \\ \hline 71 \end{array} \text{ days elapsed time}$$

Assume that a loan was made on November 20 and paid in full on January 20. Subtracting

$$\begin{array}{rl} 20 & - \quad \text{January 20} \\ -324 & - \quad \text{November 20} \\ \hline -304 \end{array}$$

yields a negative number. By adding 365 to a negative result field, we add in a full year, or 365 days, and correct thus:

$$365 + (-304) = 61 \text{ days}$$

This is the elapsed time between November 20 and January 20.

To check the number of years involved in a retention date, the input card and the table argument may also include two low-order digits containing the last two digits of the year:

$$71001 \quad \text{means} \quad \text{January 1, 1971.}$$
$$73063 \quad \text{means} \quad \text{March 4, 1973.}$$

Leap years, or years in which February contains 29 days, require additional coding by the programmer.

13-6 BINARY SEARCH

For large tables containing 100 or more entries, a more efficient technique for table searching can be used. The technique, known as *binary search,* begins searching at the middle of the table instead of at the beginning and continues to divide the remaining entries in half until the desired entry is located.

For convenience of illustration, we shall use a table of only 30 entries. (See Table 13-4.) In this table, inventory items are the search arguments and prices are the functions. Note that if our table contained 100 inventory items, an average of 50 comparisons would be required to find each desired entry. Further, table look-up for each entry (see Fig. 13-5) requires five or six computer instructions, so that in a 100 element table, searching under table look-up routines could require that six instructions be repeated a costly 100 times if we began at the top.

Table 13-4

Entry	Item #*	Price
1	01	1.25
2	02	2.00
3	03	3.45
4	06	6.98
5	08	0.65
6	10	1.58
7	12	2.76
8	15	2.50
9	17	2.85
10	19	1.90
11	21	4.15
12	22	0.95
13	28	2.13
14	40	1.04
15	42	2.16
16	45	1.81
17	51	4.45
18	60	8.00
19	63	8.50
20	70	9.00
21	71	1.05
22	79	2.18
23	80	6.43
24	83	7.04
25	88	4.02
26	90	5.14
27	91	5.58
28	94	6.35
29	98	1.23
30	99	8.50

* In ascending sequence.

To begin the search at the middle of the table, we take the first and last entry and average them:

$$\frac{1 + 30}{2} = 15\tfrac{1}{2}$$

Because no entry can be accessed by $\tfrac{1}{2}$, we truncate the fractional part of our answer and start searching at the fifteenth entry, which is the item numbered 42. Three things are possible at this point:

1. The search argument equals the table argument.
2. The search argument is lower than the table argument.
3. The search argument is higher than the table argument.

   If the search argument equals the table argument, we have found the required item. If the search argument is higher than the table argument, the required item is in the upper half of the table. If the search argument is lower than the table argument, the required item is in the lower half of the table.

   Suppose the search argument is 21.

$$\frac{1 + 30}{2} = 15\tfrac{1}{2}$$

   After truncation occurs, comparison between the search argument, 21, and the table argument, 42, reveals 21 to be in the lower half of the table since 21 < 42. (See Table 13-4.)

   Next, take the lower half of the table and divide this group of table entries in half:

$$\frac{1 + 15}{2} = 8$$

Comparing 21 to Item #15, we find that table argument to be in the second quarter of the table. Averaging gives

$$\frac{8 + 15}{2} = 11\tfrac{1}{2}$$

and truncating the $\tfrac{1}{2}$ yields 11. The search argument and the table argument are now equal, and we have found the required item 21.

Exercises 13-1

1. Complete:

	Multiplier	Multiplicand	Product
a.	xxx.xx	xxxx.xx	
b.		xxx.xx	xxxx.xxx
c.	x.xxx		xxx.xxxxxxx

2. Given 12465 ÷ 14, complete:

   a.

   b.

   c.

3. Describe the error in logic in the following Fortran statement.

      NET = RATE * HOURS + (HOURS − 40.0 * 1.5)

4.   HIGH = X(1)
   10   DO 40 I = 2, 100
   20   IF (X(I) − HIGH) 40, 40, 30
   30   HIGH = X(I)
   40   CONTINUE

   The above routine, which could be found in a Fortran program, places the highest value in an array named X into a field named HIGH. Write a routine to place the lowest number in a field named SMALL.

5. Flowchart your own routine to take care of an overflow condition during an addition operation. The procedure should include instructions to set up additional result fields. (See page 280 of this chapter.)

6. Using the binary search technique, flowchart the logic required to locate any table argument in Table 13-4.

7. Describe the rounding activity in the following report of daily deposits. Note that the columns called "Ratio to Total" and "To 1 Decimal %" do not yield the desired 100%, but that the last column, denoted "%," does.

Day	Amount Deposited	Ratio to Total	To 1 Decimal %	Rounding	%
Mon.	$ 1,062.25	.094653	9.4	.095153	9.5
Tue.	437.62	.038995	3.8	.039148	3.9
Wed.	118.00	.010515	1.0	.010663	1.0
Thur.	1083.52	.096549	9.6	.097212	9.7
Fri.	8521.15	.759289	75.9	.759501	75.9
	$11,222.54	1.000001	99.7%	1.001677	100.0

8. Design a method to determine a check digit for a seven-digit employee number.

9. The following is used to compute dates in the *Julian* Calendar. (This method is used by some computers in the Job Control Language section of programs.)

JAN = 000	JUL = 181
FEB = 031	AUG = 212
MAR = 059	SEP = 243
APR = 090	OCT = 273
MAY = 120	NOV = 304
JUN = 151	DEC = 334

To find any day of the year expressed as a three-digit number, add the day of the month to the number assigned above. For example,

May 15 expressed as a 3-digit number is 120 + 15 = 135. (For leap years add 1 more for every day after February 28.)

The year expressed as a 2-digit number is annexed to the high-order digit of the day. Thus our date in this example (May 15, 1973 is 73135. Using this method, express the following dates in five digits.

a. Oct. 21, 1974
b. April 5, 1973
c. March 1, 1972

10. Given the following Cobol instructions, describe the activities taking place. How are first and last records handled? (Assume records in DRIVER-FILE and LICENSE-FILE are sequenced in ascending order.)

```
        OPEN DRIVER-FILE. OPEN LICENSE-FILE.
        READ LICENSE-FILE.
        MOVE LICENSE-NUMBER TO SAVE.
    BEGIN.
        READ DRIVER-FILE AT END GO TO NEXT-SEARCH.
        IF PLATE-NUMBER < SAVE GO TO BEGIN.
        IF PLATE-NUMBER = SAVE GO TO FOUND-IT.
        IF PLATE NUMBER > SAVE GO TO INVALID-PLATE.
    FOUND-IT.
        WRITE FOUND-CAR-RECORD FROM LICENSE-RECORD.
        CLOSE DRIVER-FILE.
    INVALID-PLATE.
        DISPLAY "LICENSE-NUMBER OUT OF SEQUENCE OR
        ILLEGAL-PLATE-NUMBER VERIFY RECORDS".
        CLOSE DRIVER-FILE.
```

If you are familiar with Cobol, you can easily change the GO TO's to PERFORM, modify the program in a more structured manner, and compare all the PLATE-NUMBERs to LICENSE-NUMBERs.

13-7 FILE MAINTENANCE

One of the most important activities in processing information is file maintenance. For example, in order to prepare bills for customers, all transactions for purchases must be recorded within a reasonable amount of time. Typically, two files must be processed:

—A master file of all customers containing such data items as name, customer number, and balance due.
—A transaction file of purchases made within the last billing period.

When updating a master file in this manner, three events are possible (assuming records in both files are sequenced in ascending order on customer number):

1. A record on the master file contains a customer number that is less than the transaction customer number. This indicates that the master record is inactive. No purchases were made in this period, and no processing should take place for that customer.
2. The customer numbers in the master file and on the transaction record are equal. This means the customer has made purchases during the period and his balance must be updated.
3. The customer number in the master file is greater than the customer number on the transaction record. This indicates a new customer whose record must now be inserted into the master file.

As an exercise, use the flowchart below and replace the question marks with annotations that will describe the activities taking place.

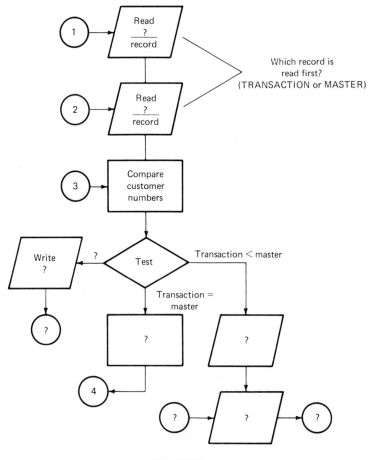

FIGURE 13-6

14

PERT AND CPM
NETWORKS

As business systems have become more complex, the process of managing such systems has become more difficult, and business firms, government agencies, and universities have expended vast amounts of money on developing techniques for use in the management process. Of these techniques, two of the most popular are PERT (Program Evaluation and Review Technique) and CPM (Critical Path Method). Both PERT and CPM are based on the concept of using a network as a model for an actual system. The basic techniques of network construction and calculations are the subjects of the following sections. We will use the building of a house as an example.

_____ 14-1 CONSTRUCTING THE NETWORK

The first step in constructing a network is to determine the individual activities which must be performed in order to achieve the desired objective. Here an *activity* is defined as any operation which requires time and/or resources and which has a definite beginning and end. Fig. 14-1 is a list of activities associated with building a house.

Once activities have been determined, a network may be constructed which graphically displays the interrelationships among activities. This graphic representation is sometimes called an *arrow diagram,* and the process of creating the diagram, *arrow diagramming.*

Identification		Duration in Days
A	Lay foundation	5
B	Erect shell	6
C	Roofing	3
D	Rough plumbing	3
E	Rough electrical	2
F	Flooring	4
.G	Windows and doors	2
H	Siding	3
I	Grading	1
J	Wallboard	4
K	Finsh plumbing	2
L	Paint interior	2
M	Finish floors	3
N	Finish electrical	2
O	Paint exterior	3
P	Landscaping	2
Q	Finishing touches	1

FIGURE 14-1

In arrow diagramming, each activity is represented by an arrow and has an origin event and a terminal event associated with it, where an *event* denotes the specific starting or ending point for an activity or group of activities and has no expenditure of resources associated with it. In diagrams, events are represented by circles and are numbered, the terminal-event number always being greater than the origin-event number. Thus, in Fig. 14-2, the arrow represents the activity, "Lay Foundation," and 1 and 2 represent the origin event ("Start of House") and terminal event ("Foundation Complete"), respectively.

FIGURE 14-2

In constructing the network, it is frequently more convenient to start with the final activity and work backward by asking which activities must be completed before it can begin. After these activities have been selected, continue to work backward in the same way. This approach is called *backward planning.*

In Fig. 14-1, activity Q is the last activity, and activities D, J, L, M, N, O, and P must be completed before it can start. Fig. 14-3 depicts what a network might look like at the initial stage of its construction using backward planning. Event numbers are chosen arbitrarily, as the number of events is not yet known. Event 14 is associated with the completion of the seven activities that must be completed before activity Q can start. That these activities may be going on concurrently is denoted by the seven arrows ending at 14.

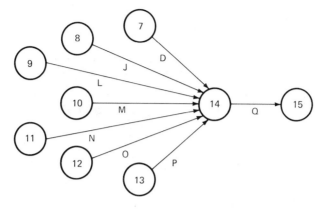

FIGURE 14-3

An alternative approach would be to start with the initial activity and work forward by asking what activities come next. This approach is called *forward planning*.

In Fig. 14-1, activity A is the initial activity and may be followed immediately by activities B and C. Fig. 14-4 depicts what the network might look like at the initial stage using forward planning. The two arrows starting at 2 represent activities B and D, which may possibly be performed concurrently and are dependent upon the completion of activity A.

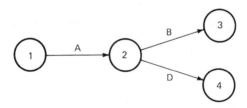

FIGURE 14-4

Backward planning is generally considered more reliable because it asks the more objective question, "What must have already occurred?" rather than the more subjective question, "What comes next?" In practice, however, a combination of both techniques is usually employed, working back and forth across the network until it is completed.

Fig. 14-5 is the completed arrow diagram for the network associated with building the house (Fig. 14-1). The dotted lines called *constraint lines* represent "dummy" or "zero-time" activities. Although no real activity is associated with going from 4 to 7, activity C must be completed before activity J can start, so this zero-time activity is necessary to depict the true relationship between C and J. Similarly, the constraint line between 12 and 15 denotes a zero-time activity that must be completed before activity Q can start.

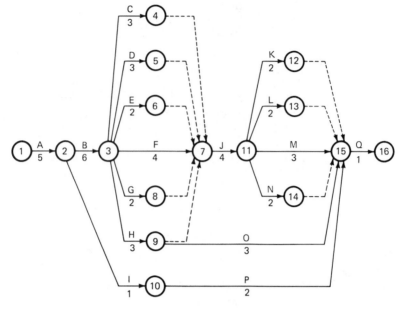

FIGURE 14-5

The network (arrow diagram) is a graphic representation of the sequence in which activities take place. As such it also represents the *plan* of action.

To use the network as a *scheduling* tool requires the addition of time estimates for the activities. Obtaining objective and accurate time estimates is usually quite difficult. However, as a time estimate is associated with an individual activity, its effect on the entire network is minimal. Sometimes, on large projects, underestimates are cancelled out by overestimates.

_____ 14-2 NETWORK CALCULATIONS

We may now use the network of Fig. 14-5, in conjunction with reasonable time estimates for each activity, to perform the calculations necessary to develop a schedule. Observe that in the figure activities I and P together require only 3 days to complete. When should they be scheduled? The arrow diagram does not tell when to schedule them, but can be used to determine

1. the earliest start date (E.S.)
2. the latest start date (L.S.)
3. the earliest completion date (E.C.)
4. the latest completion date (L.C.)

14-2.1 Forward Pass Calculations

"Earliest start dates" and "earliest completion dates" are computed by a *forward pass* through the diagram. First assume some baseline from which to start—either an actual date, such as May 15th, or some other starting time, such as "day 0." Event 1 in Fig. 14-5 is assumed to have occurred at day 0. Thus, the E.S. for the first activity is 0 and the E.C. (from the table in Fig. 14-6) is 5.

The E.C. for a given activity is used as the E.S. for the succeeding activity. Having selected a baseline, which becomes the E.S. for the first activity, compute:

$$E.C. = E.S. + Duration$$

Activity	Duration	E.S.	E.C.	L.S.	L.C.	T.S.	F.S.
A	5	0	5	0	5	0	0
B	6	5	11	5	11	0	0
C	3	11	14	12	15	1	1
D	3	11	14	12	15	1	1
E	2	11	13	13	15	2	2
F	4	11	15	11	15	0	0
G	2	11	13	13	15	2	2
H	3	11	14	12	15	1	0
I	1	5	6	19	20	14	0
J	4	15	19	15	19	0	0
K	2	19	21	20	22	1	1
L	2	19	21	20	22	1	1
M	3	19	22	19	22	0	0
N	2	19	21	20	22	1	1
O	3	14	17	19	22	5	5
P	2	6	8	20	22	14	14
Q	1	22	23	22	23	0	0

FIGURE 14-6

Thus, we have:

Activity	E.C.	=	E.S.	+	Duration
B	11	=	5	+	6
I	6	=	5	+	1

When the start of an activity is dependent on the completion of more than one preceding activity, the E.S. is the maximum E.C. of the preceding activities. For example, to calculate the E.C. and E.S. of J, compute:

Activity	E.C.	=	E.S.	+	Duration
C	14	=	11	+	3
D	14	=	11	+	3
E	13	=	11	+	2
F	15	=	11	+	4
G	13	=	11	+	2
H	14	=	11	+	3

Thus:

$$J \qquad 19 = 15 + 4$$

The E.C. for the final activity establishes the expected completion date of the project. This amount—the duration of the project time—is equivalent to the time of the longest sequence of activities in the network. This sequence is called the *critical path*. In Fig. 14-5, the critical path is:

(See Fig. 14-6 for the E.S. and E.C. for each of the activities in Fig. 14-5.)

14-2.2 Backward Pass Calculations

"Latest start dates" and "latest completion dates" are computed by starting with the last activity and working backward (by a *backward pass*) along each path to the baseline, or beginning event, to find out when the project should be initiated. First assume a scheduled completion date (usually the expected completion date of the project), then calculate

$$\text{L.S.} = \text{L.C.} - \text{Duration}$$

Thus, the L.C. for the last activity in Fig. 14-5 is 23, and the L.S. is 22.

Working backward, the L.S. of an activity is used as the L.C. for the preceding activity. Thus, we have:

Activity	L.S.	=	L.C.	−	Duration
Q	22	=	23	−	1
(12)——(15)	22	=	22	−	0
(13)——(15)	22	=	22	−	0
M	19	=	22	−	3
(14)——(15)	22	=	22	−	0
O	19	=	22	−	3
P	20	=	22	−	2

When several activities are dependent on the completion of one activity, the L.C. of that activity is the *minimum* L.S. of the succeeding activities. Hence, we can calculate the L.S. and L.C. for event J in Fig. 14-5:

Activity	L.S.	=	L.C.	−	Duration
Q	22	=	23	−	1
(12)——(15)	22	=	22	−	0
K	20	=	22	−	2
(13)==(15)	22	=	22	−	0
L	20	=	22	−	2
M	19	=	22	−	3
(14)——(15)	22	=	22	−	0
N	20	=	22	−	2

Thus:

J	15	=	19	−	4

The L.S. for the first activity establishes the latest date on which the project must start if the objectives of the network are to be met. If the L.C. for the last activity was chosen as the expected completion date calculated from the forward pass, then the L.S. for the first activity should be the baseline of the forward pass. (See Fig. 14-6 for the backward pass calculations for Fig. 14-5.)

14-2.3 Slack Calculations

In Fig. 14-6, note that the earliest start date and the latest start date for any given activity may be, but are not necessarily, the same. The difference in these dates represents the amount of flexibility available for the scheduling of that activity, and is called *total slack* (T.S.), or *total float* (T.F.). T.S. or T.F. represents the amount of time the start of an activity can be delayed without delaying the completion of the project, and can be calculated as:

$$\text{T.S.} = \text{L.C.} - \text{E.C.} \quad \text{or} \quad \text{T.S.} = \text{L.S.} - \text{E.S.}$$

Free slack (F.S.) or float is the flexibility allowed for an activity when all preceding activities start as early as possible and all succeeding activities start as early as possible. An activity with free slack may be delayed without rescheduling any other activity.

In performing the calculations for Fig. 14-6, we treated an activity and an immediately succeeding zero-time activity as though they were only one activity; in other words

were considered as one activity, or

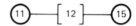

The slack calculations also indicate which activities lie in the critical path. Note that in Fig. 14-6, the activities in the critical path of Fig. 14-5 all have slack of zero.

In the example used in this chapter, the network was relatively simple and the calculations were easily done by hand. In most applications of networks, projects are more complex and calculations laborious. The calculations and associated reports are usually done by computer.

It is important to remember that the network and the associated calculations are only tools to be used in developing the schedule. The manager must also consider the availability of resources (manpower, money, space, equipment, etc). This is necessary to avoid setting unrealistic manpower requirements, overuse or underuse of equipment, delays due to undelivered materials, and so forth.

Once a plan and schedule have been prepared, the network representing them may then be used as a tool to *evaluate progress*. As activities are completed, the network is updated and analyzed. If circumstances warrant, the network may be modified to reflect more accurate time estimates. External influences such as labor strikes may also cause modifications. Changes made in the network are also reflected in corresponding modifications of the *plan* and *schedule*.

In summary, network analysis is a powerful tool for use in each step of the management process cycle. (See Fig. 14-7.)

ADVANTAGES OF PERT/CPM

1. Forces management to *plan* a project before it begins.
2. Requires an analytical approach to planning.
3. Separates the planning and scheduling functions.
4. Permits the planner to concentrate on the relationship of items of work without considering their occurrence in time.
5. Allows the planner to develop a more detailed plan, since he is concerned with *how* the work will be performed, not when.
6. Results in a more realistic schedule.
7. Clearly shows dependency relationships between work tasks.
8. Facilitates *control* of a project.
9. Simplifies maintenance of the plan and schedule.
10. Informs management of the current status of the project.
11. Focuses management attention on critical items of work.
12. Gives management the ability to assess consequences of anticipated changes to the plan.
13. Makes it easy to relate other functions of project control to the basic planning and scheduling function.
14. Meets contractual requirements of government, private industry, and customers.

FIGURE 14-7

Exercises 14-1

1. Define the terms *activity* and *event* as applied to networks. How are they represented in a network?
2. What are two ways of constructing a network?
3. Define *float* and *total float* as applied to network calculations. How are they calculated?
4. Define *critical path* and explain how it is determined.
5. Construct a network for the following:

   Project: Paint House

   Activities: Choose color
   Buy paint
   Mix paint
   Paint house
   Paint trim
   Clean brushes

6. In Fig. 14-8:

   a. Calculate for each activity

      1. earliest start time
      2. earliest complete time
      3. latest start time
      4. latest complete time
      5. total slack

   b. Determine the critical path.

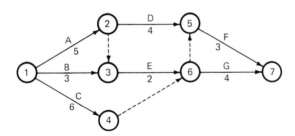

FIGURE 14-8

APPENDIX A
COMMON LOGARITHMS

x	0	1	2	3	4	5	6	7	8	9
1.0	.0000	.0043	.0086	.0128	.0170	.0212	.0253	.0294	.0334	.0374
1.1	.0414	.0453	.0492	.0531	.0569	.0607	.0645	.0682	.0719	.0755
1.2	.0792	.0828	.0864	.0899	.0934	.0969	.1004	.1038	.1072	.1106
1.3	.1139	.1173	.1206	.1239	.1271	.1303	.1335	.1367	.1399	.1430
1.4	.1461	.1492	.1523	.1553	.1584	.1614	.1644	.1673	.1703	.1732
1.5	.1761	.1790	.1818	.1847	.1875	.1903	.1931	.1959	.1987	.2014
1.6	.2041	.2068	.2095	.2122	.2148	.2175	.2201	.2227	.2253	.2279
1.7	.2304	.2330	.2355	.2380	.2405	.2430	.2455	.2480	.2504	.2529
1.8	.2553	.2577	.2601	.2625	.2648	.2672	.2695	.2718	.2742	.2765
1.9	.2788	.2810	.2833	.2856	.2878	.2900	.2923	.2945	.2967	.2989
2.0	.3010	.3032	.3054	.3075	.3096	.3118	.3139	.3160	.3181	.3201
2.1	.3222	.3243	.3263	.3284	.3304	.3324	.3345	.3365	.3385	.3404
2.2	.3424	.3444	.3464	.3483	.3502	.3522	.3541	.3560	.3579	.3598
2.3	.3617	.3636	.3655	.3674	.3692	.3711	.3729	.3747	.3766	.3784
2.4	.3802	.3820	.3838	.3856	.3874	.3892	.3909	.3927	.3945	.3962
2.5	.3979	.3997	.4014	.4031	.4048	.4065	.4082	.4099	.4116	.4133
2.6	.4150	.4166	.4183	.4200	.4216	.4232	.4249	.4265	.4281	.4298
2.7	.4314	.4330	.4346	.4362	.4378	.4393	.4409	.4425	.4440	.4456
2.8	.4472	.4487	.4502	.4518	.4533	.4548	.4564	.4579	.4594	.4609
2.9	.4624	.4639	.4654	.4669	.4683	.4698	.4713	.4728	.4742	.4757
3.0	.4771	.4786	.4800	.4814	.4829	.4843	.4857	.4871	.4886	.4900
3.1	.4914	.4928	.4942	.4955	.4969	.4983	.4997	.5011	.5024	.5038
3.2	.5051	.5065	.5079	.5092	.5105	.5119	.5132	.5145	.5159	.5172
3.3	.5185	.5198	.5211	.5224	.5237	.5250	.5263	.5276	.5289	.5302
3.4	.5315	.5328	.5340	.5353	.5366	.5378	.5391	.5403	.5416	.5428
x	0	1	2	3	4	5	6	7	8	9

x	0	1	2	3	4	5	6	7	8	9
3.5	.5441	.5453	.5465	.5478	.5490	.5502	.5514	.5527	.5539	.5551
3.6	.5563	.5575	.5587	.5599	.5611	.5623	.5635	.5647	.5658	.5670
3.7	.5682	.5694	.5705	.5717	.5729	.5740	.5752	.5763	.5775	.5786
3.8	.5798	.5809	.5821	.5832	.5843	.5855	.5866	.5877	.5888	.5899
3.9	.5911	.5922	.5933	.5944	.5955	.5966	.5977	.5988	.5999	.6010
4.0	.6021	.6031	.6042	.6053	.6064	.6075	.6085	.6096	.6107	.6117
4.1	.6128	.6138	.6149	.6160	.6170	.6180	.6191	.6201	.6212	.6222
4.2	.6232	.6243	.6253	.6263	.6274	.6284	.6294	.6304	.6314	.6325
4.3	.6335	.6345	.6355	.6365	.6375	.6385	.6395	.6405	.6415	.6425
4.4	.6435	.6444	.6454	.6464	.6474	.6484	.6493	.6503	.6513	.6522
4.5	.6532	.6542	.6551	.6561	.6571	.6580	.6590	.6599	.6609	.6618
4.6	.6628	.6637	.6646	.6656	.6665	.6675	.6684	.6693	.6702	.6712
4.7	.6721	.6730	.6739	.6749	.6758	.6767	.6776	.6785	.6794	.6803
4.8	.6812	.6821	.6830	.6839	.6848	.6857	.6866	.6875	.6884	.6893
4.9	.6902	.6911	.6920	.6928	.6937	.6946	.6955	.6964	.6972	.6981
5.0	.6990	.6998	.7007	.7016	.7024	.7033	.7042	.7050	.7059	.7067
5.1	.7076	.7084	.7093	.7101	.7110	.7118	.7126	.7135	.7143	.7152
5.2	.7160	.7168	.7177	.7185	.7193	.7202	.7210	.7218	.7226	.7235
5.3	.7243	.7251	.7259	.7267	.7275	.7284	.7292	.7300	.7308	.7316
5.4	.7324	.7332	.7340	.7348	.7356	.7364	.7372	.7380	.7388	.7396
5.5	.7404	.7412	.7419	.7427	.7435	.7443	.7451	.7459	.7466	.7474
5.6	.7482	.7490	.7497	.7505	.7513	.7520	.7528	.7536	.7543	.7551
5.7	.7559	.7566	.7574	.7582	.7589	.7597	.7604	.7612	.7619	.7627
5.8	.7634	.7642	.7649	.7657	.7664	.7672	.7679	.7686	.7694	.7701
5.9	.7709	.7716	.7723	.7731	.7738	.7745	.7752	.7760	.7767	.7774
6.0	.7782	.7789	.7796	.7803	.7810	.7818	.7825	.7832	.7839	.7846
6.1	.7853	.7860	.7868	.7875	.7882	.7889	.7896	.7903	.7910	.7917
6.2	.7924	.7931	.7938	.7945	.7952	.7959	.7966	.7973	.7980	.7987
6.3	.7993	.8000	.8007	.8014	.8021	.8028	.8035	.8041	.8048	.8055
6.4	.8062	.8069	.8075	.8082	.8089	.8096	.8102	.8109	.8116	.8122
6.5	.8129	.8136	.8142	.8149	.8156	.8162	.8169	.8176	.8182	.8189
6.6	.8195	.8202	.8209	.8215	.8222	.8228	.8235	.8241	.8248	.8254
6.7	.8261	.8267	.8274	.8280	.8287	.8293	.8299	.8306	.8312	.8319
6.8	.8325	.8331	.8338	.8344	.8351	.8357	.8363	.8370	.8376	.8382
6.9	.8388	.8395	.8401	.8407	.8414	.8420	.8426	.8432	.8439	.8445
7.0	.8451	.8457	.8463	.8470	.8476	.8482	.8488	.8494	.8500	.8506
7.1	.8513	.8519	.8525	.8531	.8537	.8543	.8549	.8555	.8561	.8567
7.2	.8573	.8579	.8585	.8591	.8597	.8603	.8609	.8615	.8621	.8627
7.3	.8633	.8639	.8645	.8651	.8657	.8663	.8669	.8675	.8681	.8686
7.4	.8692	.8698	.8704	.8710	.8716	.8722	.8727	.8733	.8739	.8745
7.5	.8751	.8756	.8762	.8768	.8774	.8779	.8785	.8791	.8797	.8802
7.6	.8808	.8814	.8820	.8825	.8831	.8837	.8842	.8848	.8854	.8859
7.7	.8865	.8871	.8876	.8882	.8887	.8893	.8899	.8904	.8910	.8915
7.8	.8921	.8927	.8932	.8938	.8943	.8949	.8954	.8960	.8965	.8971
7.9	.8976	.8982	.8987	.8993	.8998	.9004	.9009	.9015	.9020	.9025
8.0	.9031	.9036	.9042	.9047	.9053	.9058	.9063	.9069	.9074	.9079
8.1	.9085	.9090	.9096	.9101	.9106	.9112	.9117	.9122	.9128	.9133
8.2	.9138	.9143	.9149	.9154	.9159	.9165	.9170	.9175	.9180	.9186
8.3	.9191	.9196	.9201	.9206	.9212	.9217	.9222	.9227	.9232	.9238
8.4	.9243	.9248	.9253	.9258	.9263	.9269	.9274	.9279	.9284	.9289
8.5	.9294	.9299	.9304	.9309	.9315	.9320	.9325	.9330	.9335	.9340
8.6	.9345	.9350	.9355	.9360	.9365	.9370	.9375	.9380	.9385	.9390
x	0	1	2	3	4	5	6	7	8	9

x	0	1	2	3	4	5	6	7	8	9
8.7	.9395	.9400	.9405	.9410	.9415	.9420	.9425	.9430	.9435	.9440
8.8	.9445	.9450	.9455	.9460	.9465	.9469	.9474	.9479	.9484	.9489
8.9	.9494	.9499	.9504	.9509	.9513	.9518	.9523	.9528	.9533	.9538
9.0	.9542	.9547	.9552	.9557	.9562	.9566	.9571	.9576	.9581	.9586
9.1	.9590	.9595	.9600	.9605	.9609	.9614	.9619	.9624	.9628	.9633
9.2	.9638	.9643	.9647	.9652	.9657	.9661	.9666	.9671	.9675	.9680
9.3	.9685	.9689	.9694	.9699	.9703	.9708	.9713	.9717	.9722	.9727
9.4	.9731	.9736	.9741	.9745	.9750	.9754	.9759	.9763	.9768	.9773
9.5	.9777	.9782	.9786	.9791	.9795	.9800	.9805	.9809	.9814	.9818
9.6	.9823	.9827	.9832	.9836	.9841	.9845	.9850	.9854	.9859	.9863
9.7	.9868	.9872	.9877	.9881	.9886	.9890	.9894	.9899	.9903	.9908
9.8	.9912	.9917	.9921	.9926	.9930	.9934	.9939	.9943	.9948	.9952
9.9	.9956	.9961	.9965	.9969	.9974	.9978	.9983	.9987	.9991	.9996
x	0	1	2	3	4	5	6	7	8	9

APPENDIX B
ALGEBRA REVIEW

This appendix should be used as a *reference* for definitions and a review of operations on powers and roots. Most of the material is at the high school mathematics level. (The section dealing with powers and roots is presented in a manner that facilitiates computer programming of the material.)

DEFINITIONS

Definition B-1

A *constant* is a number having a single value for a particular problem, such as 4, −32, 0.071.

Definition B-2

A *term* consists of constants, variables, or combinations of either or both. Constants or variables are combined by multiplication in a term. (Variables are described on page 2.)

EXAMPLES $3s$, $4y$, $2x$, abc, -8^2y^3.

Definition B-3

An *expression* consists of constants or variables combined by any of the arithmetic operators. An expression must consist of at least one constant.

EXAMPLES $2x - 1$, $x^2 + 3x + C$, $(x + 1)(x - 1)$, $\frac{2}{3}x^3$, $1x + y$.

Definition B-4

An *open sentence* is a sentence which is neither true nor false; that is, it cannot be determined if it is true or false from the given information. Open sentences must have at least one variable.

EXAMPLES

a. She is a cheerleader. Since we do not know who *she* is, the statement is neither true nor false.
b. $2 + 9 = y$

If we have no value for y, the statement is an open sentence.

Definition B-5

A *closed sentence* is a sentence which is true or false, but not both.

EXAMPLES

a. Roger Staubach was a quarterback (true).
b. $3 < 0.05$ (false).
c. $3 + 2 = 5$ (true).

Definition B-6

An *equation* is an open sentence or closed sentence in which two quantities are equal.

EXAMPLES

a. $4 + 5 = 9$
b. $3x + 3 = 11$

Definition B-7

The *replacement set* of a sentence is a set whose members may be used to replace a variable in an open sentence to make a closed sentence.

EXAMPLE   Given the open sentence $2x + 3 = 5$ and the replacement set $I = \{. . . -2, -1, 0, 1, 2 . . .\}$, then $2x + 3 = 5$ can be made true only when $x = 1$. Any other members of the replacement set will make $2x + 3 = 5$ false.

Definition B-8

A *solution* is an element from the replacement set which makes the sentence true.

Definition B-9

The *solution set* of a statement is the set whose members are all solutions of some open sentence.

EXAMPLES

a. In $2x + 5 = 8$, the solution set is $\{\frac{3}{2}\}$.
b. In $x^2 - 2x = 0$, the solution set is $\{0, 2\}$.
c. In $2x + 5 > 8$, the solution set is $\{x \mid x = 2, 3, 4 \ldots\}$ if the replacement set contains positive integers only. A conditional equation is an equation which has a solution set. In $2x + 2 = 8$, the solution set is $\{3\}$, which contains one element.

Equivalent equations have the same solution set. For example,

$$2a + 3 = 13$$
$$a + 7 = 12$$

are equivalent equations whose solution set is $\{5\}$.

An *identity* is an equation whose solution set must contain the set of all elements of the replacement set. For example,

$$7x - 3x = 4x$$

is an identity because any element in the set of real numbers may replace the variable x.

If the solution set contains no members, it can be represented as $\varnothing$, the null set, and the equation is a false sentence. $y = y - 3$ is an example of a false sentence.

Exercises B-1

1. Give two equations equivalent to $2x + 4 = 12$.
2. Give an example of

    a. A conditional equation
    b. An identity

3. Find the solution set for

    a. $3x + 9 = 29$
    b. $2x = 2x + 1$
    c. $x^2 + 3x - 4 = 0$
    d. $2x = y$

4. Find the values of $2x + 3$ for the replacement set

$$\{-1, 0, 1, 2, 3\}$$

5. Find the values of $x - 4$ for the replacement set

$$\{-6, -4, -2, 0, 2, 4, 6\}$$

6. Find the solution set for the equation $|3x + 8| = 11$.
7. a. List the elements in the replacement set

$$A = \{x \in J | -4 < x < 5\}$$

   b. Find the solution set for the inequality $|x| > 2$.
8. Find the solution set for
   a. $|x| > 9$
   b. $|x| < 9$

EXPONENTS

Definition B-10

When two or more numbers are multiplied together, each one of them is called a factor. If all factors are equal to each other, the number of factors is called a *power* of that factor.

For example, the factors $a \cdot a$ may be written as a^2, $a \cdot a \cdot a \cdot a$ can be written as a^4, and in general, we write a^n, where a is called the base and n is called the power or exponent of the base. The term a^4 is described as "a to the fourth power"; a^n is called "a to the nth power."

Some general statements about exponents are:

1. Let a equal any real number; then $a^1 = a$ and $a^1 \cdot a^1 \cdot a^1 = a^3$.
2. $a^n = 1$ for $a = 1$ and any n.
3. 0^0 is undefined.
4. $a^{-n} = \dfrac{1}{a^n}$

Following are some of the basic laws that govern the use of exponents for any real numbers a, b, m, and n with a, $b > 0$.

1. $a^m \cdot a^n = a^{m+n}$
2. $a^m \cdot b^m = (ab)^m$
3. $a^m/a^n = a^{m-n}$ if $m > n$ and $a \neq 0$
4. $a^m/a^n = 1$ if $m = n$ and $a \neq 0$
5. $a^m/a^n = \dfrac{1}{a^{n-m}}$ if $m < n$ and $a \neq 0$
6. $a^m/b^m = \left(\dfrac{a}{b}\right)^m$ if $b \neq 0$
7. $(a^m)^n = a^{mn}$

It is not the purpose of this text to prove the laws of exponents. However, the following examples should be studied because they demonstrate the use of these laws.

1. $a^m \cdot a^n = a^{m+n}$
   Let $m = 3, n = 4$.
   Then $a^3 \cdot a^4 = (a \cdot a \cdot a) \cdot (a \cdot a \cdot a \cdot a)$
   $\qquad = a^{3+4}$
   $\qquad = a^7$

2. $a^m \cdot b^m = (a \cdot b)^m$
   Let $a = 4, b = 3, m = 2$.
   Then $4^2 \cdot 3^2 = (4 \cdot 3)^2$
   $\qquad\qquad = 144$ (Likewise, $16 \cdot 9 = 144$)

3. $a^m/a^n = a^{m-n}$ for $a \neq 0$ and all $m, n \in R$
   Let $m = 5, n = 3$ $(m > n)$.
   Then $\dfrac{a \cdot a \cdot a \cdot a \cdot a}{a \cdot a \cdot a}$ means $a^{5-3} = a^2$

4. $a^m/a^n = 1$ when $m = n$ and $a \neq 0$. If $m = n = 4$,
   then $\dfrac{a^1 \cdot a^1 \cdot a^1 \cdot a^1}{a^1 \cdot a^1 \cdot a^1 \cdot a^1} = 1$
   or $a^{4-4} = a^0 = 1$

5. $a^m \div a^n = \dfrac{1}{a^{n-m}}$ if $m < n$, $a \neq 0$, and $\dfrac{1}{a^{n-m}} = a^{-(n-m)}$
   Let $a = 10, m = 2, n = 3$.
   Then $\dfrac{10 \cdot 10}{10 \cdot 10 \cdot 10} = \dfrac{1}{10^{3-2}}$ or $\dfrac{1}{10} = 10^{-1}$

6. $a^m \div b^m = \left(\dfrac{a}{b}\right)^m$ if $b \neq 0$
   Let $a = 6, b = 4, m = 3$.
   Then $6^3 \div 4^3 = \left(\dfrac{6}{4}\right)^3$
   $\qquad\qquad = \dfrac{27}{8}$
   Likewise, $\dfrac{6^3}{4^3} = \dfrac{216}{64}$
   $\qquad\qquad = \dfrac{27}{8}$

7. $(a^m)^n = a^{mn}$
   Let $m = 3, n = 4$.
   Then $(a \cdot a \cdot a)^4 = (a \cdot a \cdot a) \cdot (a \cdot a \cdot a) \cdot (a \cdot a \cdot a) \cdot (a \cdot a \cdot a)$
   $\qquad\qquad = a^{3 \cdot 4}$
   $\qquad\qquad = a^{12}$

Exercises B-2

Perform the following operations.

1. $x^5 \cdot x^6$
2. $a^3 \cdot a^4 \cdot (2a)^2$
3. $y^4 \cdot z^4$
4. $a^6 \div a^2$
5. $a^x \div b^x$
6. $x(12^3 \div 6^3)$
7. $3^4 \div 3^6$
8. $5(4^2 \div 4^2)$
9. $2y \cdot (y^3)^2$
10. $8^2 \cdot 16^3 \cdot 64 = 2?$
11. $a^{-1} =$
12. $3^4 \cdot 9^2 \cdot 3^{-1} \cdot 27 = 3?$
13. When $a = -1$, what is a^n when n is even? When n is odd?

14. $\dfrac{x^n}{x^{-n}}$

15. $\dfrac{x^{n-1}}{x^n} \div \dfrac{x^{-n}}{x^{n+1}}$.

RADICALS

Definition B-11

"The nth root of x is y," denoted as $\sqrt[n]{x} = y$, if and only if $y^n = x$.

EXAMPLE   If $n = 2$ and $x = 4$, then $y = \sqrt{4} = \pm 2$ since $(\pm 2)^2 = 4$. We describe $\sqrt{4}$ as the second, or square, root of 4.

If $n = 3$ and $x = -8$, then $y = \sqrt[3]{-8} = -2$ since $(-2)^3 = -8$. We describe $\sqrt[3]{-8}$ as the cube root of -8.

A radical is any indicated root. The symbol that denotes a root to be taken is $\sqrt{}$. If any root other than 2 is to be taken, the number specifying that root is written in the radical sign ($\sqrt[3]{27}$) and is called the *index*. The number under the radical sign is called the *radicand*. The nth root of a radicand x can be expressed as $\sqrt[n]{x}$.

Some roots are elements in the rational set of numbers, for example,

$$\sqrt[4]{16} = \pm 2, \text{ since } (\pm 2)^4 = 16$$
$$\sqrt{a^2 \cdot b^4} = \pm(ab^2), \text{ since } (\pm ab^2)^2 = a^2 b^4$$
$$\sqrt[3]{1/8} = \tfrac{1}{2}, \text{ since } (\tfrac{1}{2})^3 = \tfrac{1}{8}$$

If the root cannot be represented by an integer or a rational (fractional) number, it is called an irrational number.

EXAMPLES $\qquad$ $\sqrt{2}, \sqrt[3]{100}$

Other roots, such as the root for $\sqrt{-1}$, are not typical of data processing problems and are not discussed in this text. (See also pg. 3.)

There are four basic laws that govern operations on radicals:

1. $\sqrt[n]{x^n} = (\sqrt[n]{x})^n = x$, for $\sqrt[n]{x} \in R^*$
   states that the order in finding the power or the root is not important.
   That is, $\sqrt[3]{27^3} = (\sqrt[3]{27})^3 = 27$
2. $\sqrt[n]{a} \cdot \sqrt[n]{b} = \sqrt[n]{a \cdot b}$ for $\sqrt[n]{a}$ and $\sqrt[n]{b} \in R$
   states that radicals with the same index can be multiplied.
3. $\sqrt[n]{a} \div \sqrt[n]{b} = \sqrt[n]{a/b}$ and $\sqrt[n]{a}$ and $\sqrt[n]{b} \in R$
   states that radicals with the same index can be divided.
4. $x\sqrt{a} + y\sqrt{a} = (x + y)\sqrt{a}$ for $x\sqrt{a}$ and $y\sqrt{a} \in R$
   states that radicals that have the same index and radicand can be added or subtracted. Thus,

$$2\sqrt{a} + 3\sqrt{a} = 5\sqrt{a}$$
$$2\sqrt[3]{x} - 3\sqrt[3]{x} = -\sqrt[3]{x}$$

For convenience in manual operations, as well as in computer programming, it is always helpful to simplify a radical expression to one that can be easily operated on. The following transformations are frequently used:

T.1. *Use as small a radicand as possible.*
In $\sqrt{250}$ we can use our second law for radicals:

$$\sqrt{250} = \sqrt{25 \cdot 10} = \sqrt{25} \cdot \sqrt{10} = 5\sqrt{10}$$

T.2. *Use a rational denominator.*
In $3/\sqrt[3]{9}$ we can use the property of the multiplicative identity and the second law for radicals:

$$\frac{3}{\sqrt[3]{9}} \times \frac{\sqrt[3]{81}}{\sqrt[3]{81}} = \frac{3\sqrt[3]{81}}{\sqrt[3]{9 \cdot 81}}$$
$$= \frac{9\sqrt[3]{3}}{9}$$
$$= \sqrt[3]{3}$$

T.3. *Use an element from the set of integers as a radicand.*
In $\sqrt{\frac{3}{7}}$ we can again use the property found on page 5 and the third law for radicals:

$$\sqrt{\tfrac{3}{7} \cdot 1} = \sqrt{\tfrac{3}{7} \cdot \tfrac{7}{7}}$$
$$= \sqrt{\tfrac{21}{49}}$$
$$= \sqrt{\tfrac{1}{49}} \cdot \sqrt{21}$$
$$= \tfrac{1}{7}\sqrt{21}$$

---

* Note that $\sqrt[2]{-4^2} = (\sqrt[2]{-4})^2$, but $\sqrt[2]{-4} \notin R$, where R represents the set of real numbers.

At times, it may be desirable to reverse procedures found in transformation 1. (See T.1.)

T.4. *Use a coefficient of 1 or 2.*

In $5\sqrt{5}$

we can apply the first law for radicals thus:

$$5\sqrt{5} = \sqrt{5^2} \cdot \sqrt{5}$$
$$= \sqrt{25} \cdot \sqrt{5}$$
$$= \sqrt{125} \text{ or } 1\sqrt{125}$$

Before continuing with transformations of radicals, we note that the notions of exponents and integral roots are related in Definition B-11, and that the exponents we have used so far have had integral values. We can also use nonintegral exponents (and also those which are irrational).

Consider Definition B-11 and $\sqrt[3]{8} = 2$ (since $2^3 = 8$). For $\sqrt[3]{8} = 2$, we can write equivalently $8^{1/3} = 2$ and express the third root as the fractional exponent $\frac{1}{3}$.

Understanding that a number in radical form is another way of expressing the same number in exponential form makes it possible to write the *eighth law of exponents.*

8. $\sqrt[n]{a^m} = (\sqrt[n]{a})^m = a^{m/n}$ for $\sqrt[n]{a} \in R$ and $m, n \in R$.

EXAMPLE   Let

$$a = 16, m = 4, n = 2.$$

Then

$$\sqrt[2]{16^4} = (\sqrt[2]{16})^4$$
$$= 16^{4/2}$$
$$= 256$$

Likewise,

$$\sqrt{65536} = 256$$

and

$$\sqrt[2]{(256)^2} = 256$$

Using the eighth law of exponents, we can now complete our methods of simplifying radicals.

T.5. *Use as low an index as possible.*

EXAMPLE 1
$$\sqrt[6]{16} = \sqrt[6]{2^4}$$
$$= (\sqrt[6]{2})^4$$
$$= 2^{4/6}$$
$$= 2^{2/3}$$
$$= \sqrt[3]{2^2}$$
$$= \sqrt[3]{4}$$

EXAMPLE 2

$$\sqrt[6]{64} = \sqrt[6]{2^6}$$
$$= (\sqrt[6]{2})^6$$
$$= 2^{6/6}$$
$$= 2$$

Exercises B-3

Perform the indicated operations.

1. $3\sqrt[3]{27}$
2. $4^2 + 2\sqrt{64} + 3\sqrt{64}$
3. $\sqrt{3} - 2\sqrt{3}$
4. $2^5 + \sqrt{2} + 16^{3/2} + 16^{3/2} \cdot 4^0$
5. $9\sqrt{5} + 6\sqrt{125}$
6. $4\sqrt{4} + 3\sqrt{16} + \sqrt{128}$
7. $\dfrac{3}{2\sqrt{3}} + \dfrac{7}{\sqrt{3}} - \dfrac{6}{4\sqrt{3}}$
8. $\sqrt{490} + \sqrt{250} + x\sqrt{360}$
9. $\sqrt{x^4}$
10. $\sqrt[x]{4^2}$
11. Using the law for fractional (rational) exponents, show that $\sqrt[n]{a} \cdot \sqrt[n]{b} = \sqrt[n]{ab}$, where $a \geq 0$, $b \geq 0$.

Simplify the following:

12. $\sqrt{72}$
13. $1/3\sqrt{45}$
14. $\sqrt{5/8}$
15. $\dfrac{5}{\sqrt[4]{16}}$
16. $\sqrt[3]{81^2}$
17. $\sqrt[8]{625}$
18. $\sqrt{x} \cdot \sqrt{xy}$
19. $\sqrt[4]{16xy^4}$
20. $\left(\dfrac{9}{64}\right)^{3/2}$
21. $\left(\dfrac{8}{125}\right)^{2/3}$
22. $\sqrt{\dfrac{10}{4}}$
23. $\sqrt[5]{12} \cdot \sqrt[5]{8}$

24. $\sqrt{\dfrac{2x}{6y}}$

25. $\sqrt{\dfrac{a + b}{a^2b^2}}$

26. $\dfrac{\sqrt{ab}}{\sqrt{ac}}$

27. $\dfrac{\sqrt[3]{ab^2}}{\sqrt[3]{a^2b^3}}$

In general, if $a \neq 0$, $a^{-n} = \dfrac{1}{a^n}$ for n any positive integer. Where exponents are fractional (rational) and negative, take the reciprocal and proceed as before.

EXAMPLE $\qquad\qquad \left(\dfrac{1}{27}\right)^{-2/3} = (27)^{2/3} = 9$

Perform the following:

28. $\left(\dfrac{8}{125}\right)^{-2/3}$

29. $\dfrac{36^{-1/2} \cdot 9^{-1/2}}{81^{1/4}}$

30. $\left[4\sqrt[4]{\dfrac{x^8}{16}}\right]^{-1/2}$

31. $(\sqrt{x^{-3}})^{-2/3}$

To find the solution set S for $\sqrt{x - 1} = 6$, square both sides:
$$(\sqrt{x - 1})^2 = 6^2$$
$$x - 1 = 36$$
$$x = 37$$
$$S = \{37\}$$

Find the solution set for problems 32-34.

32. $\sqrt{x^2 + 3} = 8 + x$

33. $2\sqrt{x + 4} - x = 1$

34. $\sqrt[3]{x + 1} = 3$

35. Complete, using the $>$, $<$, or $=$ signs, for $x > 1$.

   a. If $b > 1$, then b^x __ 1.
   b. If $0 < b < 1$, then 0 __ b^x __ 1.
   c. If $b^{1/x} = a$, then a^x __ b.

36. Do exercises a, b, and c describe properties of $f(x) = b^x$? ($f(x)$ is described on page 187.)

THE QUADRATIC FORMULA

To solve $ax^2 + bx + c = 0$ for x, we can express the solution set of a quadratic formula in terms of its coefficients when $a \neq 0$.

Step 1. Subtract c from each side of the equation.

$$ax^2 + bx = -c$$

Step 2. Divide by a.

$$x^2 + \frac{bx}{a} = -\frac{c}{a}$$

Step 3. Complete the square by adding $\left(\frac{b}{2a}\right)^2$ to both sides of the equation.

$$x^2 + \frac{bx}{a} + \frac{b^2}{4a^2} = -\frac{c}{a} + \frac{b^2}{4a^2}$$

Step 4. Find the square of the left side of the equation.

$$\left(x + \frac{b}{2a}\right)^2 = \frac{b^2}{4a^2} - \frac{c}{a}$$

Step 5. Express the right side of the equation as one fraction.

$$\left(x + \frac{b}{2a}\right)^2 = \frac{b^2 - 4ac}{4a^2}$$

Step 6. Take the square root of both sides.

$$\left(x + \frac{b}{2a}\right) = \pm \sqrt{\frac{b^2 - 4ac}{4a^2}}$$

$$x + \frac{b}{2a} = \pm \frac{\sqrt{b^2 - 4ac}}{2a}$$

Step 7. Subtract $b/2a$ from both sides of the equation to find x.

$$x = -\frac{b}{2a} \pm \frac{\sqrt{b^2 - 4ac}}{2a}$$

Step 8. Expressing the right side as one fraction, we obtain the quadratic formula.

$$x = \frac{-b \pm \sqrt{b^2 - 4ac}}{2a}$$

INDEX